SYSTEM ERROR

WHERE BIG TECH
WENT WRONG AND
HOW WE CAN REBOOT

• ■ • ■ •

ROB REICH,
MEHRAN SAHAMI,
AND JEREMY M. WEINSTEIN

HARPER

An Imprint of HarperCollins*Publishers*

To our amazing children

HarperCollins books may be purchased for educational, business, or sales promotional use. For information, please email the Special Markets Department at SPsales@harpercollins.com.

FIRST EDITION

Library of Congress Cataloging-in-Publication Data has been applied for.

ISBN 978-0-06-306488-1

21 22 23 24 25   LSC   10 9 8 7 6 5 4 3 2 1

CONTENTS

PART I: Decoding the Technologists

CHAPTER 1: The Imperfections of the
Optimization Mindset

CHAPTER 2: The Problematic Marriage of
Hackers and Venture Capitalists

CHAPTER 3: The Winner-Take-All Race Between
Disruption and Democracy

PART II: Disaggregating the Technologies

PART III: Recoding the Future

PREFACE

In times of crisis, ordinary citizens, confused and disoriented,
settling into paralysis, can come to believe that, as Plato had
argued, they are not up to the job of making difficult decisions.
In hard times, democratic citizens may become more willing
to hand over the business of politics to experts and to abandon
the institutional frameworks, the rights and liberties, that
secure their position as participants in the political process.
The danger of intellectual paralysis in the face of chaos is
finally that it undermines the first premise of democracy:
namely, that ordinary citizens will always be ready to think.

—Danielle Allen, Aims of Education Address,
September 20, 2001

On January 6, 2021, the US Capitol was stormed by insurrectionists who
had been whipped into action at a rally earlier that day featuring President Donald Trump. Their goal was to violently overturn the result of the
presidential election, which they had been falsely told for weeks had been
"stolen." That message had been delivered most prominently by Trump
himself, despite more than sixty failed lawsuits challenging election results and thorough refutations by election officials across the country.

The big tech platforms had been a key conduit for the accusations
of election fraud for months before. On January 6, they finally woke up
to the horror they had enabled. Twitter locked Trump's account, which
had nearly 90 million followers, denying him permission to post. Two
days later, citing a "risk of further incitement of violence," Twitter permanently banished Trump from the platform, erasing everything on his
account in one fell swoop. Similar suspensions took place on Facebook,

Instagram, YouTube, and Snapchat. Trump turned to the still active @POTUS Twitter account and posted that he had been "SILENCED!" before that tweet was quickly removed by the platform as well.

The platforms' rebuke of Trump's election disinformation was also an alarm bell about how much power is concentrated in the hands of a few big tech companies. The president of the United States—often touted as the "leader of the free world"—was unceremoniously stripped of his favorite means of communicating with his tens of millions of followers. Whether that was a necessary step to reduce the possibility of further violence after the election, a long-overdue decision by the platforms to take away a megaphone from a man whose history of lies far pre-dated the 2020 election, or the tech elite's blatant censorship of the highest elected official in the United States, it cast an unmistakable light on the extraordinary power that technology and, more significantly, the people who develop it have over us.

Big tech's role in and reaction to the events that led to the storming of the US Capitol only highlight the concerns about technology that have been mounting for years. Seemingly endless reports of privacy breaches and stories of behavior manipulation resulting from vast troves of data mined by large companies have made it commonplace to view big tech through a dark lens. Some argue that the internet, smartphones, and computers have delivered to us a set of devices hell-bent on hijacking our attention and addicting us to the screen, while gathering ever more data of our online behavior. And as borne out at the Capitol, a tidal wave of misinformation and disinformation on social media platforms has served to undermine our trust in science, exacerbate political polarization, and threaten democracy itself—all of this powered by a small number of companies with immense market power and growing political influence.

At this same unprecedented time, we experienced the COVID-19 pandemic, which as of this writing has taken more than 3 million lives worldwide while upending work, education, the economy, and our personal lives. The pandemic caused one of those rare moments of instantaneous behavior change with extraordinary long-term implications. Vladimir Lenin is alleged to have said, "There are decades when nothing happens, and then there are weeks where decades happen." Overnight, much of the world shifted to working from home and schools closed as public health authorities imposed social distancing rules and in some areas shelter-in-place orders. Videoconferencing soared as air travel ground to a halt. Technologies for file sharing and workplace collaboration enabled many aspects of the economy to proceed apace. People flocked in record numbers to Netflix as a substitute for movie theaters. The use of Facebook and other social media networks skyrocketed as people sought connections to friends and family. Videoconferencing enabled children to keep attending school and people to retain a connection to their loved ones when it wasn't possible to be together physically. And tech companies across the board stepped up to foreground authoritative scientific information about the pandemic, develop contact-tracing apps to help contain it, and deploy artificial intelligence to hasten the development of medical treatments and potential vaccines and to power robots to handle tasks such as delivering medication to sick hospital patients.

In short, our professional and personal lives, our economy and intimate relationships, and even our health would have been far worse without the internet and our familiar addictive devices.

As we exit the COVID-19 pandemic and enter a new political moment, the window is finally opening for a mature consideration of technology, one that avoids both the technoboosterism that accompanied its early decades and the "techlash" that has followed.

Sure, there remain plenty of criticisms to be made of Facebook, the privacy policies of Zoom, the acceleration of automation in an age of smart machines without regard for job displacement, and the toxic misinformation and disinformation flowing through social media platforms. But that just underscores the essential work of our new post-pandemic era. We must strive now to find ways to harness the power of technology to deliver its considerable benefits while diminishing its equally apparent harms to individuals and societies. We now possess the wisdom to see technological innovation as something other than an external force that works upon us. The path of technological development and the effects of technology on us are things we can shape. Things we must shape.

When we uncritically celebrate technology or unthinkingly criticize it, the end result is to leave technologists in charge of our future. This book was written to provide an understanding of how we as individuals, and especially together as citizens in a democracy, can exercise our agency, reinvigorate our democracy, and direct the digital revolution to serve our best interests.

● ■ ● ■ ●

For the past twenty years, we have been teaching at Stanford University, the seedbed of Silicon Valley. It is a research powerhouse with numerous Nobel laureates, MacArthur Foundation geniuses, and Pulitzer Prize–winning writers to rival the best. But behind the façade of this paradisiacal, self-professed "nerd nation," we started to observe some concerning patterns.

Innovation and *disruption* were the buzzwords on campus, and our students broadcasted an almost utopian view that the old ways of doing things were broken and technology was the all-powerful solution: it could end poverty, fix racism, equalize opportunity, strengthen democracy, and even help topple authoritarian regimes. "Every year at new-student orientation," one of our students told us enthusiastically, "we bring in some tech billionaire who is held up as the paragon of what you can achieve and that that's the life you should want." The former president of the university was heard to say that government was incompetent and the idea of encouraging any student to go into government service in order to make a difference was "ridiculous."

Perhaps most disconcerting, the enthusiasm for the digital economy and the moneymaking pipeline from Stanford to Silicon Valley was not tempered by critical reflection on just whose problems were being solved (and whose were ignored), who was benefiting from innovation (and who was losing), and who had a voice (and who remained unheard) in shaping our technological future.

This is not just a Stanford point of view. Many of the same pathologies we've identified are on display on a broader scale. For example, even with the blowback against technologists, uncritical headlines around the globe too often claim that technology will solve our most complex problems, whether climate change, poverty, or mental health crises—a naive optimism we have worked hard to counter in our students. "Making the world a better place" has become more a punch line than a real mission statement for major technology companies, underscoring the difficulty many of us face in determining what is truly in the public interest.

We joined forces to try to bring about a cultural intervention on campus that might reverberate into the tech world and beyond. Our view was simple: we cannot find a path to a better technological future without the three distinct perspectives we bring to the table.

Mehran Sahami was recruited to Google in its start-up days by Sergey Brin. One of the inventors of email spam filtering technology, Mehran spent a decade in the industry working on applications that are now used by billions of people. In 2007, with a background in machine learning and AI, he returned to Stanford as a computer science professor; he wants technologists to understand that the decisions they make in producing code have real social consequences that affect millions of people. Though engineers may write code with good intentions, Mehran is concerned that, too often, social consequences are not considered until a major screw-up makes the problem transparent for everyone. At that point, it may be too late.

Jeremy Weinstein went to Washington with President Barack Obama in 2009. A key staffer in the White House, he foresaw how new technologies might remake the relationship between governments and citizens, and launched Obama's Open Government Partnership, a global network of governments, NGOs, and technologists fighting to ensure that governments deliver for people. He then joined Samantha Power in New York when she was appointed US Ambassador to the United Nations. In the wake of North Korea's cyberattack on Sony and the FBI-Apple encryption fight, they confronted the enormous gulf between those who build technology and those who bear the responsibility for governing a society transformed by technology. But just as policy makers are ignorant of technology in many ways, technologists are naive about and perhaps even willfully blind to the importance of public policy and the ways that social science can help us understand, anticipate, and even mitigate the impacts of technology on society. When he returned to Stanford in 2015 as a professor of political science, Jeremy made it his top priority to teach young computer scientists and to bring social science to the study of how technologies are reshaping our social environments.

Rob Reich is a philosopher who is a leader of the university's Center for Ethics in Society and Institute for Human-Centered Artificial Intelligence. He brings a Socratic orientation, asking probing and uncomfortable questions designed to shake up the perspective of the technologist: What makes disruption valuable? Why obsess over optimization? Is increasing click-through rates on digital advertisements your highest calling? Perhaps most important, he wants to challenge engineers' perception of their role. It's not enough to be a problem solver without asking deeper questions: Is this a problem worth solving? Are there particular ways we should solve it given the things we value? Given the power of technology, who deserves a seat at the table in defining the problems and seeking solutions? Where does democracy fit in, if at all?

We brought our collective expertise together and designed a new course on the ethics and politics of technological change that quickly became one of the most popular classes on campus. While our three perspectives—the technologist, the policymaker, and the philosopher—are central to the course, we recognized that other voices were also essential. In our teaching, we sought to incorporate perspectives on technology that go beyond our own: communities of color that are disproportionately harmed by particular innovations, those whose livelihoods might be threatened by automation, women shining a light on the sexist culture in tech, and activists fighting the power of the C-suite from both inside and outside of the companies. People off campus started asking us to bring the material to a bigger audience, first with a public version open to hundreds of community members and later with an evening class for engineers, entrepreneurs, and venture capitalists in San Francisco.

In each setting, we found that people were primed for a discussion that could get beyond the scandal du jour, move past the enthusiasts and the polemicists, and start grappling with what it means to tackle these issues head-on. Students were struggling with what it means to pursue a career in technology at a moment when the harmful impacts of new technologies can no longer be ignored. Professionals were asking hard questions about whether it is possible to reform tech companies from within. And for those outside of the tech sector, there was a clear desire to take stock of the power of big tech and reckon with their own sense of powerlessness to shape its direction.

Though it was not a surprise to see that these issues were salient, we watched people struggle to articulate the values that they felt were at stake with each new innovation and to take a stand in defense of those values, especially if it came at some cost in terms of efficiency, convenience, or profit. It's hard enough to justify our most important values and to understand how societies have sought to defend and preserve them. But it is even more difficult to determine how value trade-offs should be handled or whether they can be addressed in any systematic way.

With this book, we hope to engage you—as a person who uses or works with technology and as a citizen who has so much at stake   in thinking through a new path forward.

INTRODUCTION

Joshua Browder entered Stanford as a young, brilliant undergraduate in 2015. His Wikipedia page describes him as a "British-American entrepreneur," and he's already been named to *Forbes* magazine's "30 Under 30" list. As a freshman at Stanford—after no more than three months there, he says—he programmed a chatbot to help people overturn their parking tickets. He'd thought of the start-up when he was living in London before college: "I got thirty parking tickets in the UK when I was in high school at about eighteen years old, the driving age. I couldn't pay for any of the tickets. I probably deserved them, but because I couldn't afford them, I created software for myself and my friends to get out of them." Seems simple enough for a side project during your first year of college, but of course Browder discovered that "everyone in the world hates parking tickets." Fast-forward a few years, and Browder was on leave from Stanford as the CEO of a tech company called DoNotPay, which provides a free and automated mechanism for challenging parking tickets issued in big cities, including London and New York. According to a glowing profile of his work, as of June 2016, the company had successfully challenged more than 160,000 parking tickets, sparing people $4 million.

The service is pretty straightforward. Browder worked with a group of pro bono traffic lawyers to identify the most common reasons for parking tickets to be overturned. A chatbot asks users a few questions that enable it to make a judgment about whether the user can file an effective appeal. The chatbot then guides the user through the process of filing an appeal, at no charge. The chatbot has little capacity to determine whether a ticket was issued legitimately; it simply provides the user with the optimal grievance procedure. Obviously, users are thrilled to get out of paying annoying and often expensive parking tickets, and the only people who lose are lawyers and the government. In Browder's words, "parking tickets are a sort of tax on the vulnerable. It's so wrong that the government is taxing the group they should be protecting." Browder has accordingly been celebrated as a "wunderkind" in magazines and websites such as *Wired*, *Business Insider*, and *Newsweek*, as well as at Stanford itself. And he's secured the support of one of Silicon Valley's most successful venture capital firms, Andreessen Horowitz, which led the seed round of funding for Browder's company in 2017.

But this is exactly the type of story—and there are hundreds of them at Stanford and in Silicon Valley—that gives us pause. From our perspective, it's essential to reflect on why parking tickets exist in the first place. Annoying as they might be, they serve many important, legitimate purposes. They deter people from parking by fire hydrants, blocking driveways, or occupying spaces reserved for the disabled. In large cities, they motivate people to move their cars for street cleaners. Enhanced parking enforcement can also be used to achieve broader community priorities, such as reducing traffic and congestion. And parking tickets constitute a meaningful source of municipal revenue necessary to support a city and its citizens.

Browder may have been responding to a zeitgeist in the conservative London tabloids that slammed local governments' efforts to raise revenues through parking tickets, something that coincided with other city initiatives to reduce traffic and congestion for reasons of both convenience and environmental health. But reducing traffic is something that a lot of people just might value. In London, local councils must spend the revenue from parking tickets on local transport projects, including the £9 billion backlog in national road repairs. Infrastructure is a classic example of a public good—difficult for the market to supply because, in the absence of government intervention, consumers will take advantage of the infrastructure without paying the costs of using it. Hence there is a role for taxes, fines, and, yes, parking tickets. As for whether parking tickets are a tax on the vulnerable, there actually aren't any good data that reveal who pays parking tickets. But in a city with as efficient and affordable a public transportation system as London's, it's fair to assume that low-income families are much more likely than the upper class to ride buses and the Tube. Once one digs a little beneath the surface, the argument that parking tickets are a tax on the vulnerable doesn't sound too convincing.

The story becomes even more worrisome when one asks Browder about his broader ambitions. After all, in Silicon Valley, the CEO of a successful start-up is always considering how to further scale up the company. "I would like to hopefully replace lawyers with technology," he says, "starting with very simple things like arguing against parking tickets and then moving toward things like pressing a button and suing someone or pressing a button and getting a divorce." Browder's long-term vision is that you'll never need a trained, human lawyer again and that "consumers won't even know what the word lawyer means." This is probably music to the ears of many who detest the legal profession, bemoan our society's litigiousness, and are envious of lawyers' salaries, which might seem outsize relative to their societal role and contribution. But do we really want to live in a society where people can sue at the push of a button? Would divorce be less painful if algorithms and automated systems were making decisions about who should have custody of the kids and how shared property should be divided?

We don't want to single out Browder's pursuit as particularly malignant. He is not a bad person. He just lives in a world where it is normal not to think twice about how new technology companies could create harmful effects. Browder is but one recent example of the start-up mindset birthed at Stanford and in Silicon Valley at large. He's been encouraged by his professors, his peers, and his investors to think bigger and be ambitious. But too rarely do people stop and ask: Whose problem are you solving? Is it a problem actually worth solving? And is the solution proposed one that would be good for human beings and for society?

Back in 2004, just as Silicon Valley was reemerging from the "dot-com bust," a young man named Aaron Swartz enrolled at Stanford University.

Like Browder, he had been fascinated by computer programming from an early age. He'd won a national prize at the age of thirteen for his creation of an online collaborative library, theinfo.org. At fourteen, he helped create the Really Simple Syndication (RSS) specification, a widely used internet protocol that permitted automatic access to updates on websites anywhere. The goal was to create open standards that would allow anyone to share and update information on the internet.

Swartz enrolled straightaway in an accelerated course on computer programming while also taking introductory classes in sociology, a seminar on Noam Chomsky, and a required first-year humanities class on freedom, equality, and difference. He found Stanford to be alienating, however. In an online daily journal he kept for a few weeks, he recorded his dissatisfaction with his fellow students—too shallow—and his courses. The humanities lecture, he wrote, "turns out to consist mostly of the three professors arguing with each other about what a paragraph really means. . . . Is this what the humanities is like? Even the RSS debates were better than this."

Swartz spent much of his time coding on his own. During his freshman year, he applied to join Y Combinator, a newly created tech incubator, to start a company called Infogami that would help manage content on websites. He was selected for the very first cohort of Y Combinator's Summer Founders Program. By the end of the summer, he decided to continue working on the company, which would soon merge with another Y Combinator start-up, Reddit. Two years later Reddit was sold to Condé Nast, reportedly for between $10 million and $20 million, and Swartz became a young millionaire. Reddit is today one of the most popular sites on the internet and is valued at $3 billion.

A brilliant young coder goes to college, then drops out to pursue his start-up dreams. Sounds like the same kind of dropout story that was told about Bill Gates and Steve Jobs and would be told again about Mark Zuckerberg and Elizabeth Holmes; the same story that Joshua Browder is currently living out.

But Aaron Swartz was different. He was less interested in making money than in using technology to change how human beings access and interact with information. "Information is power," he wrote in a "Guerrilla Open Access Manifesto" in 2008, "but like all power, there are those who want to keep it for themselves. . . . But you need not—indeed morally you cannot—keep this privilege for yourselves. You have a duty to share it with the world."

As a fifteen-year-old, before even entering Stanford, Swartz emailed one of the world's leading tech intellectuals, Lawrence Lessig, to ask if he could help in writing the code for what would become Creative Commons, a system of online copyright licenses that permit people to use, share, and modify creative work without cost. Swartz viewed technology as inextricably bound up with politics and saw the effort to control information as a way to control people. He wanted a liberatory technology because he thought it would help bring about a liberatory politics.

The language of freedom, equality, and justice was interspersed in his lexicon with the language of coding and internet protocols. His views about technology made him a technology activist. His views about how technology was connected to politics made him a political activist. The two went hand in hand, and his activism took multiple forms.

In 2008, he founded Watchdog.net, an effort to aggregate information about politicians in order to increase political transparency and stimulate grassroots activism. He contributed to the development of Open Library, which catalogues books online. In 2010, he founded the Web activist group Demand Progress, which would successfully mobilize resistance to US legislation that would undermine net neutrality. He made available to the public millions of US court records that had been archived in a digital system called Public Access to Court Electronic Records (PACER). He consistently sought to find civic and political uses for technology, and he despaired whenever technology was hijacked by coders who sought to make themselves rich without considering the effects of their technology on the world.

In 2006, he attended an international gathering of the Wikipedia community, the people who administer and contribute to the famous open-access, nonprofit, user-generated internet encyclopedia. "At most 'technology' conferences I've been to, the participants generally talk about technology for its own sake. If *use* ever gets discussed, it's only about using it to make vast sums of money." At the Wikipedia conference, however, "the primary concern was doing the most good for the world, with technology as the tool to help us get there. It was an incredible gust of fresh air, one that knocked me off my feet."

One of his other efforts was to press for open access to knowledge produced by scholars. It irritated him that in order to read the contents of online journals you either had to be a student or an employee of a university, or you had to pay considerable fees—and this despite the fact that public funds actually financed the work of scholars at both public and private universities. Why should journal articles be copyrighted, with the financial benefits flowing not to the authors of the articles but to the large corporations that owned the scientific journals? In 2010, he began downloading thousands of academic articles from a scholarly repository called JSTOR. He did so by using the computer network at MIT, where a long-standing policy of maintaining an open campus gave permission to anyone on campus, visitors included, to access its network. He wrote a program on his laptop that would automate the downloading process rather than accessing articles one by one, which was the requirement under JSTOR's terms of service. After several visits to a computer closet, where he connected his laptop to the MIT network, Swartz had downloaded millions of articles, violating JSTOR's policy and implicating MIT's network in the violation.

MIT traced the downloads to Swartz's laptop and the closet from which his computer had accessed the network, and when he came back for another round of downloads in early 2011, he was arrested by MIT police and charged with breaking and entering with intent to commit a felony. JSTOR decided to drop the charges against him after Swartz returned the data files, but MIT elected to continue with its prosecution. In 2012, federal prosecutors added nine felony counts to the charges against him, with a maximum sentence of fifty years in jail. Swartz sank into a depression, and in the midst of multiple efforts at plea bargaining and preparing to go to trial, he committed suicide in his Brooklyn apartment in early 2013. He was twenty-six years old.

It was a devastating end to a life of enormous promise, a life that had already reached celebrity status in tech circles. In the month following his death, the hackers known as Anonymous infiltrated the websites of MIT and the US State Department and declared, "Aaron Swartz this is for you." Lawrence Lessig eulogized Swartz as someone he had mentored but who, in the end, had really mentored him. Memorials sprang up around the world.

It's impossible to know what Swartz was thinking when he repeatedly violated JSTOR's terms of service. Or what prosecutors were thinking when they pressed their case even after JSTOR withdrew. And of course it's impossible to peer into the mind of a person struggling with depression and wonder what might have brought him to contemplate suicide and then to take his own life. For us, however, Aaron Swartz's death is a hinge event in the evolution of the politics and ethics of technology. His life, and what became of the world of technology after his death, illustrate broader lessons about what a technologist might bring to the world. For Swartz, learning how to code was part of amassing a tool kit for civic and political change. He was the dropout who saw technology not as a means of becoming rich but as a lever for the pursuit of justice.

While Swartz was alive, he was a hero to many and a celebrity in the world of technology: the kid who helped develop Creative Commons, the tech activist who led a movement to protect net neutrality and beat back the US Congress, the evangelist for open access to knowledge. He was the latest in successive generations of technologists who felt that technology was a tool for human empowerment and espoused unapologetically utopian and radically democratic visions of a technological future, a vision with deep roots in the creation of the internet and the culture of Silicon Valley.

Today, fewer than ten years after his death, virtually nobody talks about Aaron Swartz. He is mostly forgotten in Silicon Valley, and he is unknown to the wider public. At Stanford University, we rarely meet students who know Swartz's name or can describe what he did. They do know the names of Gates, Jobs, Zuckerberg, and former Stanford students such as Larry Page and Sergey Brin (the cofounders of Google),

Evan Spiegel and Bobby Murphy (the cofounders of Snapchat), Kevin Systrom and Mike Krieger (the cofounders of Instagram), and Elon Musk (the founder of Tesla and SpaceX). And many students on campus today know the name Joshua Browder. If they haven't heard of his successfully funded start-up, they know of his work because he spammed the entire student body in early 2019 to offer them a chance, by using his service DoNotPay, to get out of fees that support a wide array of student groups on campus.

Today, the heroic figures are the disruptive and instantly wealthy innovators. Whereas once technologists brought with them counter-cultural visions of enhancing human capabilities, promoting liberty and equality, and spreading democracy, today the culture of Silicon Valley is about founder worship and the celebration of apolitical coders. This was a profound shift that technologists didn't notice or didn't want to acknowledge until they had to in the wake of the social and political fallout from technology's role in Brexit, the election of Trump, and the siege of the US Capitol.

• ■ • ■ •

The rise of the Joshua Browders and the decline of the Aaron Swartzes encapsulate the challenge the world confronts with Silicon Valley. One of the most far-reaching transformations of our age is the wave of digital technologies rolling over and upending nearly every aspect of life. Work and leisure, family and friendship, community and citizenship—all have been reshaped by our now-ubiquitous digital tools and platforms. We know that we are at a turning point. How to think about what should be done, and why, is what we need to grapple with.

The bloom is off the rose of the big tech companies. We no longer hear so much gushing about the internet as a tool for putting a library into everyone's hands, social media as a means of empowering people to challenge their governments, or tech innovators who make our lives better by disrupting old industries. The conversation has shifted to the other pole. Humans are being replaced by machines, and the future of work is uncertain. Private companies surveil in ways that governments never even contemplated and profit handsomely in the process. The internet ecosystem feeds hate and intolerance with its echo chambers and filter bubbles. The conclusion seems inescapable: our technological future is grim.

However, we must resist this temptation to think in extremes. Both techno-utopianism and -dystopianism are all too facile and simplistic outlooks for our complex age. Instead of taking the easy way out or throwing our hands up in the air, we must rise to the defining challenge of our era: harnessing technological progress to serve rather than subvert the interests of individuals and societies. This task is not one for technologists alone but for all of us.

Tackling this challenge begins with recognizing that new technologies create civic and social by-products, or, in the language of economics, *externalities*. An externality is the cost or benefit that a person's or corporation's activity imposes on others. For example, a chemical plant that dumps its waste into a nearby river and expects others to pick up the costs is creating a negative externality. Big, unregulated tech companies that harvest our private data and sell them to the highest bidder are not that different from chemical plants; it's just the type of dumping that is different.

Facebook's business model is to increase the time we spend on its platform and then sell access to our personalized profiles to advertisers and political operatives who seek to manipulate our behavior and dump the by-product of that manipulation onto our personal lives and democratic institutions. YouTube's recommendation systems and default autoplay setting keep users watching videos on its platform while pushing people into echo chambers and feeding them more extreme content, thereby undermining our democracies, which rely on facts and trust. And Uber's and Waymo's push for automated vehicles may increase productivity but leave displaced and unemployed workers at the mercy of the government's feeble social safety net.

These by-products are not accidental but a reflection of the choices technologists make when they design and launch new products. Most of these choices are invisible to us, even though they directly impact our democracy and the well-being of our fellow citizens.

The technologists have some powerful enablers. When you marry the engineers' mindset to the venture capitalists' profit motive, you get an obsession with scale. Great companies are those that, in the words of a cofounder of LinkedIn, Reid Hoffman, "scale from zero to a gazillion"— even though scale is what can quickly turn manageable consequences into unmanageable, toxic messes. And if you add into the mix the government's failure to regulate much of anything related to new technologies, it's easy to see why we now find ourselves in an increasingly dangerous and strikingly unequal world.

The upshot is the common conviction that technological progress is rolling over us with the wheels of inevitability. Ordinary people can't undo technological discoveries and innovations, and they can't shape the products we buy or the effects of technology across society. What can a truck driver whose job may be replaced by autonomous vehicles do? What can a parent do about the apps a child is transfixed by, short of taking the phone out of the child's hands? What can an employee do about the deployment of facial recognition at the office? What can a citizen do about the spread of disinformation on the platforms that deliver us news and information? And what can anyone do about the routine collection of personal data that we accept as the price we pay for using digital gadgets and services?

Though we so often feel powerless in the face of the ubiquitous technologies in our lives, the passive acceptance of these consequences need not be the only path forward. In *System Error*, we show that the effects of technology are neither preordained nor fixed in stone. They depend on how we design new technologies, how we interact with them, and what rules we set in place to govern them.

We are entering a new era in which the rules that govern technology will no longer be written by hackers or companies alone. They will reflect some push and pull among the companies that make things, the governments that oversee them, the consumers who use them, and the people who are affected by them. As we enter this new moment, regardless of our professional role, we are consumers of these technologies. Even more important, all of us are citizens. And as citizens, each of us has a vital role to play. We need to understand what values are at stake, how new technologies create tensions between some values and ignore others entirely, and how we can most effectively shape the impact of new technologies on our society.

In this book, we distill more than three decades of experience at ground zero of the technological revolution, Stanford University and Silicon Valley, to show how each of us can play an essential role in shaping our technological future. That begins by moving beyond an obsession with particular technologies or tech companies and directing our attention instead to the distinctive mindset and growing power of the technologist. And that mindset is one of optimization.

In revealing that mindset, we confront the core issue: that well-intentioned optimizers fail to measure all that is meaningful, and when their creative disruptions achieve great scale, they impose their values and decisions upon the rest of us. A better strategy would replace the blinkered technocratic governance by coders and powerful tech companies with the messy, inefficient, yet empowering process of deciding what values to promote through what we call democracy. The story of technology can no longer be a Manichean tale of good and bad people or good and bad technologies but must be a mature reckoning with the realization that the powerful technologies dominating our lives encode within them a set of values that we had no role in choosing and that we often do not even see.

• ■ • ■ •

In this book, we will be asking a series of questions about ethics and technology—or, more specifically, about ethics and *technologists*. Our aim is to understand how the pursuit of optimization by technologists—often seen as a good in itself—can undermine the well-being of individuals and the health of democratic societies.

Ethical concerns in technology come in three basic camps. First is what we can call the problem of personal ethics. Engineers should strive to be persons of good character. For example, they shouldn't cheat, lie, or steal.

Of course there's nothing in this camp that is unique to technologists or the profession of engineering. All people, in all professions, should strive to be persons of good character.

Take the infamous case of Elizabeth Holmes, a Stanford dropout who at nineteen years of age founded Theranos, a biomedical start-up company that claimed to have engineered a revolutionary blood-testing technology. The new technology, it was claimed, could automate blood testing using pinprick amounts of blood to check for multiple conditions, allowing such testing to be done in local pharmacies or at home. At one point in 2013, Theranos was valued at more than $10 billion. It turned out, however, that Theranos had no such revolutionary technology and the enterprise was a house of cards. The fraudulent representations of its technology were exposed when some young whistleblower employees reported their concerns to state officials and investigative journalists at the *Wall Street Journal*. Holmes is currently facing criminal charges, and the company was shuttered in 2018.

Such stories exist in every profession: the deceptions of the financier Bernie Madoff; the doping by the bicyclist Lance Armstrong; the corruption of politicians such as Richard Nixon.

The issue of personal ethics is often a real problem, because some people do in fact sometimes conduct themselves, steer their businesses, or lead public agencies in an unethical manner. But these are actually the *least interesting* ethical questions that arise. No one believes that lying, cheating, or stealing is defensible behavior. The ethical lessons are not much different from those you would find in the book *All I Really Need to Know I Learned in Kindergarten*.

The second camp of ethical concerns moves beyond questions of personal character to issues of professional ethics. What professional standards should guide the behaviors and actions of individuals in a particular profession? The medical profession has the Hippocratic Oath: "First, do no harm." But more than aspirational oaths, professional ethics often involve organizational bodies that are capable of effectively policing the conduct of members of the profession, with real teeth and consequences in cases where people or companies breach a code of conduct.

Over the course of time and prodded by various scandals, medical research and practice have been subject to significant standards of professional conduct in order to protect the interests of research subjects and patients. One such scandal was the forty-year-long Tuskegee experiment. That study involved the provision of free medical care to several hundred African Americans, some of whom had already contracted syphilis and others who had been deliberately infected with it in order to study the disease. The men did not know they had the disease, and they were purposely left untreated even though the medical intervention to cure it—penicillin—was well known. The Tuskegee experiment was leaked by a whistleblower and led to numerous reforms, including the creation of institutional review boards for any research at universities or in pharmaceutical companies that involves human subjects.

The profession of computer scientist, though having developed a professional code of ethics, does not provide significant consequences for its violations. In 2018, the Association for Computing Machinery (ACM), the world's largest educational and scientific computing society with nearly 100,000 members, updated its Code of Ethics and Professional Conduct for the first time since 1992. The code outlines general ethical principles, professional responsibilities, and leadership principles that should guide the conduct of ACM members. The principles it espouses are admirable. Violation of the code is grounds for expulsion from the ACM. But in reality the threat of expulsion has no teeth: unlike state bar associations or medical board associations, the ACM is not a gatekeeper to technological professions. In fact, there is no gatekeeper. Expulsion from the ACM is largely inconsequential for those working in the industry.

The third camp of ethical concern is social and political ethics. This involves questions about public policy, regulation, and governance. The most interesting—and the most difficult—ethical challenges fall into this third category, and these are the challenges that will occupy us most frequently in this book.

Social and political ethics involve confronting and resolving trade-offs between rival values that we all acknowledge we care about. The issue here is not about discerning the difference between right and wrong and learning how to do the right thing; it is about identifying the multiple good things we care about and deciding what to do when they cannot be simultaneously realized in full.

A classic example is the tension in any society between liberty and equality. As the British philosopher Isaiah Berlin put it, total liberty for the wolves is certain death to the lambs. Though both are worthy aims, it may be impossible to fully realize them simultaneously. We have to make choices. There are trade-offs.

This is where ethics gets interesting. It is not simply a matter of teaching ethics to technologists. No required ethics course has ever successfully inoculated people, or an industry, from bad behavior. It's simply not realistic to rely on all people being moral saints when confronted with ordinary, much less difficult, decisions. As James Madison wrote in the Federalist Papers, "If Men were angels, no government would be necessary." When the stakes are high and the power of technology over our lives is enormous, relying on technologists alone, even ethical ones, is a mistake.

When confronted with rival values, the benefits of democracy—despite its flaws and limitations—come into view. Democracies aspire to give voice to all citizens and to handle persistent disagreement and competing interests. The give-and-take of democratic politics is a time-honored system for making decisions about conflicting ideas and needs. The particular advantage of democracies is that they tend to decide things slowly, through deliberation and with the standing possibility of revising any past decision.

Although democracies can strive to achieve the best outcomes for citizens, they can also serve as guardrails against *the worst outcomes*. Karl Popper, the twentieth-century Austrian philosopher, believed that the central problem of politics is not to decide who should rule: the people (a democracy), the wisest (a philosopher king or a beneficent technocrat), or the wealthiest (an oligarchy). The central problem of politics is how to organize political institutions so that terrible outcomes can be avoided and bad or incompetent rulers can be prevented from doing grave damage. This is the moment we find ourselves in as the impacts of technology overwhelm our society.

• ■ • ■ •

If we accept that technology is simply beyond our control, we cede our future to engineers, corporate leaders, and venture capitalists. Some might pin their hopes on the market, thinking that it will look out for our interests, deliver us the technologies we want, and weed out those that are not useful or might even do harm. But the market is good at some things and not at others. It rewards profit without regard to social consequences. It prizes efficiency while ignoring other values. It celebrates domination. These priorities are encoded in the algorithms that power new technologies, the metrics that drive company strategy, and the regulatory environment that governs what companies may and may not do.

But there are other ideals at stake: fairness, privacy, autonomy, equality, democracy, justice. As human beings, we treasure these concepts and protect them through the rules we've established to govern ourselves. Yet new technologies have put many of them at risk—sometimes in ways that are visible but often in ways that are not.

For the past three decades of the digital economy, it's been far too easy to rely on beneficent technologists to shape our collective trajectory. But with their moral shortcomings on full display and heightened concerns about the power of authoritarians to bend technology to serve their own interests, we must forge a different path. And the manner in which we resolve these difficult tensions must be democratic, with the active participation of citizens like you.

We can't leave our technological future to engineers, venture capitalists, and politicians. This book lays out the dangers of leaving the optimizers in charge and empowers all of us to make the difficult decisions that will determine how technology transforms our society. There are few more important tasks before us in the twenty-first century. When we act collectively, we not only take charge of our own destiny, we also make it far likelier that our technological future will be one in which individuals will flourish alongside, and because of, a reinvigorated democracy.

PART I

DECODING THE TECHNOLOGISTS

In an age of advanced technology, inefficiency
is the sin against the Holy Ghost.

—Aldous Huxley, foreword to *Brave New World*, 1946

CHAPTER 1

THE IMPERFECTIONS OF THE OPTIMIZATION MINDSET

Contrary to current popular opinion, for most of its storied existence the United States Postal Service has been a hub of disruptive innovation. In 1792, Benjamin Rush, a signer of the Declaration of Independence, ushered the Postal Service Act into law, giving the federal government control over various regional postal routes and declaring that the content of mail was private, even if delivered by a public carrier. It also provided for a novel cost structure: rather than charging on the basis of weight, the postal service would charge one rate for letters and another much less expensive rate for newspapers. The exchange and broad dissemination of news and information would thus be subsidized.

In order to improve the efficiency and reliability of mail delivery throughout a quickly expanding nation, the post office regularly turned to new technologies. It introduced delivery by horse, the Pony Express, in 1862 and began experimenting with delivery by train a few years later. To connect as many people as possible, the post office introduced free delivery of mail to rural areas in 1902. And just a few years after Henry Ford invented the automobile in 1901 and the Wright brothers the airplane in 1903, the post office was trying out both technologies for mail delivery.

In 1913, during the height of railroad development and the dawn of the age of the automobile, the post office introduced another radical and innovative idea: parcel post, designed to facilitate the delivery of ordinary merchandise via mail. In the expanding age of commerce, the post office would deliver goods in addition to letters, newspapers, and magazines. New forms of business sprang up to take advantage of this new delivery mechanism. Mail-order companies boomed. With thousands of carriers spread across the country, including in rural areas, mail-order industry profits went from $40 million in 1908 to $250 million in 1920, a huge sum at the time.

Imagine a family in 1920 in a rural area of the United States that is planning a trip and needs to buy a bunch of supplies. Rather than having to go many miles to the nearest city by horse or train, the family can order by mail from the Sears, Roebuck catalogue, choosing from an array of goods much larger than any shop in their own town is likely to stock. After receiving the order, Sears, Roebuck will fulfill it using parcel post. The process from start to finish might take a few weeks. That is still a long time but much more efficient and convenient than the alternatives.

Fast-forward a century. The post office is facing a crisis in the digital age. Digital technologies have driven greater and greater efficiencies in the exchange of messages. Email and text messaging now permit the instantaneous delivery of personal messages across the greatest of distances. The volume of single letters sent by first-class mail declined by 61 percent from 1995 to 2013. With the ubiquity of email and text messaging, writing a letter has become more and more a rarity, an expression of some romantic inclination rather than the humdrum desire to communicate or get something done. Ordering merchandise from catalogues via mail still exists but has declined precipitously. The Sears, Roebuck catalogue folded in 1993, and the company filed for bankruptcy in 2018.

How would that family in a rural area outfit themselves for a trip today? They would likely order what they need from Amazon Prime with the click of a few buttons, and in two days the packages would be delivered, possibly by the post office or a private carrier such as FedEx. In some cities, the packages might even arrive in two hours with Amazon Prime Now, bypassing the post office altogether. And Amazon Prime Air promises the future possibility of delivery by drones in thirty minutes or less. A family could shop in the morning and be ready to head out with all their supplies in the afternoon.

What a story of progress through greater and greater efficiency! Instantaneous mail! Delivery of packages within a few days, hours, or even minutes!

The evolution of Netflix tells a similar story about the drive for efficiency. At its start, Netflix was a subscription service for movies, not unlike Blockbuster, the bricks-and-mortar behemoth of the movie rental business. For a monthly fee, Netflix would deliver DVDs using the US Postal Service, providing a custom-designed, instantly recognizable red postage-paid envelope for returning them. In the company's early days, customer satisfaction hinged on there being as quick a turnaround time as possible between the return of a DVD and the arrival in the mail of the next movie in the customer's queue. If customers had to wait a week for a new movie to arrive, they complained. Although Netflix made extraordinary efforts to make the turnaround time as short as possible, it wasn't responsible for carrying the red envelopes in the mail; that was the post office's job. Netflix provided customized machines in thousands of post offices to speed up the return process, aiming for one-day delivery to its subscribers. And at one point, looking for every possible means to improve on delivery and return times, it hired the former US postmaster general to be its chief operations officer. Who better to improve Netflix's ability to navigate the postal system?

But even in its early days, Netflix's long-term plan wasn't just about mailing DVDs. Reed Hastings, the company's founder and CEO, who holds a master's degree in computer science, knew that it was only a matter of time until efficiency gains in internet communication would enable the streaming of video directly to consumers. In 2007, his vision came to fruition. Taking advantage of broadband access, Netflix could bypass the postal service altogether. The company, recognizing that it still needed to support customers with limited or no internet access, also spun out a separate service called Qwikster that would continue DVD delivery by mail. Quickly it pivoted almost entirely to delivering films and television shows on demand via streaming video. Now it's possible to watch nearly any movie one wants to at any moment, instantaneously, via the internet. Blockbuster—which in 2000 had turned down an offer to acquire Netflix for $50 million—filed for bankruptcy in 2010.

Today the delivery of movies is not merely efficient; it has been *optimized*. In terms of time spent, there appear to be no possible further improvements after instantaneous delivery via streaming.

Efficiency gains have led not only to greater convenience but to outcomes that matter far more, including many that enhance democracy or economic opportunity: better distribution of essential medicines around the globe, the development of new vaccines, easier access to the world's information, and more effective interventions for students with learning differences.

Over the past several decades, the drive toward efficiency and optimization has come to play an increasingly dominant role across spheres and industries, from business (e.g., streamlining supply chains) to sports (e.g., *Moneyball*-style tactics using big-data analytics to drive decisions) and even to our personal lives (e.g., online dating apps and fitness trackers). It's no coincidence that the industry and skill set most ascendant over the same time period has been computer science. In the digital age, the disruptive innovators tend to be the efficiency-obsessed tribe of people called coders, or software engineers. They are the ones who invent and bring to market the host of new technologies that are driving efficiency gains across so many aspects of life.

SHOULD WE OPTIMIZE EVERYTHING?

Efficiency is not always the good thing it seems to be. Consider the creation of a product called Soylent, another Silicon Valley innovation driven by engineers.

Soylent is a nutritional powder that can be made into a drinkable shake by adding water. This meal replacement product was developed because its inventor believed that food is a pain point in our daily lives, an inefficient delivery mechanism for the human body's nutritional needs. Eating food is costly and requires shopping, cooking, and cleaning up or else going out to restaurants. And many meals are social affairs, with attendant expectations of conversation and social etiquette. All of this takes considerable time away from other potentially valuable activities, such as working.

Soylent is the brainchild of Rob Rhinehart, a Silicon Valley engineer who set out to solve a specific problem. After working at a failed start-up in San Francisco, he and his friends found themselves running short on cash and facing the challenge of making a decent meal. "I started wondering why something as simple and important as food was still so inefficient, given how streamlined and optimized other modern things are," he told *Vice*. As he wrote on his personal blog, "In my own life I resented the time, money, and effort the purchase, preparation, consumption, and clean-up of food was consuming. I am pretty young, generally in good health, and remain physically and mentally active. I don't want to lose weight. I want to maintain it and spend less energy getting energy."

So he did what engineers do: he took an engineering approach to food and nutrition. He researched the vitamins and nutrients needed for bodily sustenance. He studied materials from the Food and Drug Administration and textbooks on nutrition and biochemistry, and he made a list of more than thirty nutrients required by the human body. He ordered the nutrients over the internet and began experimenting by blending them together in a powder form.

In 2013, he started living on the powder, which he would blend into a shake, making adjustments along the way when he realized that his concoction lacked iron and that he had miscalculated the amount of fiber he needed to feel healthy. After a month of subsisting on his powdered solution alone, he wrote a blog post entitled "How I Stopped Eating Food."

Rhinehart called his invention Soylent, taking his inspiration from a 1966 novel depicting an overpopulated world with dwindling resources in which a new food, made from soy and lentils (soy + lent), is created in order to feed people. For most folks, however, the name evokes the 1973 film adaptation of the novel, *Soylent Green* starring Charlton Heston. In the movie, people live on a wafer they believe to be made from plankton. The wafer is revealed in the final scenes to be produced from human flesh, and the dystopia of overpopulation turns out to be an even greater horror in which cannibalism is the only way to survive. Rhinehart never claimed to be a branding genius.

Despite this, his blog post attracted attention. It was especially popular on a site called Hacker News, a place for the tech community to learn about clever inventions and gizmos to make life better and save time. Rhinehart saw an entrepreneurial opportunity, and he posted about Soylent on a crowdfunding site, promising to deliver a week's worth of Soylent in return for a modest donation of $65. He hoped to raise $100,000 to bootstrap production. The response was enormous. He hit his target in just two hours. Eventually, more than six thousand people sent money to support his new venture, raising a total of more than $750,000.

Rhinehart and his collaborators went into business, and in 2014 they introduced Soylent to the public. The company is funded by some of Silicon Valley's most prominent venture capital firms. In an introductory video posted on YouTube, Rhinehart explained, "What I really learned was how to break problems down. Everything is made of parts, everything can be broken down. Unlike most other foods which prioritize taste and texture, Soylent was engineered to maximize nutrition; to nourish the body in the most efficient way possible."

It wasn't just the desire to maximize his nutritional needs that motivated him. He told a reporter that farms where food is grown and animals are cultivated are "very inefficient factories." "It's really the labor that gets me," he said. "Agriculture's one of the most dangerous and dirty jobs out there, and it's traditionally done by the underclass. There's so much walking and manual labor, counting and measuring. Surely it should be automated."

Soylent, he said, would solve multiple problems: the inefficiency of feeding oneself with food, the stress of worrying about optimal nutrition, the food industry's reliance on farms. Soylent would deliver all that at relatively low cost. Win-win-win-win.

The press covered the release of Soylent as "The End of Food." In the *New York Times*, Farhad Manjoo complained about its "stultifying utilitarianism," calling it a "punishingly boring, joyless product." Soylent may offer complete nourishment but only, the tech columnist wrote, "at the expense of the aesthetic and emotional pleasures many of us crave in food."

The newspaper assigned Sam Sifton, its food critic and restaurant reviewer, to give Soylent a try. The results were predictable: "Imagine a meal made of the milk left in the bottom of a bowl of cut-rate cereal, the liquid thickened with sweepings from the floor of a health food store, and you have some sense of what it is like to consume the protein-packed shakes that have replaced Flamin' Hot Cheetos and Red Bull in the diets of some tech workers in Silicon Valley." Summoning up a small bit of charity, he concluded, "These instant meals are meant for work warriors for whom good and delicious food is secondary to perfect and unassailable engineering."

It's not hard to spot some problems with Soylent. It may be a hyperefficient means of meeting one's daily nutritional needs while drastically cutting down on the time required to prepare and eat regular food. But for most people, food is not just a delivery mechanism for one's nutritional requirements. Food serves many different ends. It brings gustatory pleasure. It provides social connection. It sustains and transmits cultural identity. A world in which Soylent spells the end of food also spells the loss of these values.

Maybe you don't care about Soylent; it's just another product in the marketplace that no one is required to buy. If tech workers want to economize on time spent grocery shopping or a busy person faces the choice between grabbing an unhealthy meal at a fast-food joint or bringing along some Soylent, why should anyone complain? In fact, it's a welcome alternative for some people.

That's fine. Engineering has obviously brought humanity lots of good. But the story of Soylent is powerful because it reveals the optimization mindset of the technologist. And problems arise when this mindset begins to dominate—when the technologies begin to scale and become universal and unavoidable.

THE EDUCATION OF AN ENGINEER

In 1936, John Maynard Keynes, one of the most influential economists ever to live, observed the following:

> *The ideas of economists and political philosophers, both when they are right and when they are wrong, are more powerful than is commonly understood. Indeed the world is ruled by little else. Practical men, who believe themselves to be quite exempt from any intellectual influence, are usually the slaves of some defunct economist. Madmen in authority, who hear voices in the air, are distilling their frenzy from some academic scribbler of a few years back.*

He wrote those words in the early part of the twentieth century, at a moment of extreme global upheaval, with the Great Depression giving way to a second world war. One might not have thought that ideas were what mattered most at such a moment. But he was right. The perspectives of economists and the contest of political ideologies served to shape the two world wars, the Cold War, the fall of the Berlin Wall, the rise of the financial sector, and a globalizing economy—virtually all of the greatest challenges of the twentieth century.

Economists also entered the innermost halls of political decision-making, advising leaders and directly crafting public policy. Prior to World War II, it was lawyers who dominated federal agencies, and courts ignored most economic evidence about the predicted effects of their decisions. But economists flooded into public service in the mid–twentieth century, growing their ranks from about two thousand in the federal government in the 1950s to more than six thousand in the 1970s. They were hired in droves by large companies to drive growth and provide economic forecasts. And the leaders of the booming Wall Street banks, private equity firms, and hedge funds that rose to prominence in the last quarter of the century all had a background in economics.

What economics and finance were to the twentieth century, engineering and computer science are to the twenty-first. Computer hardware, processing power, big data, algorithms, artificial intelligence (AI), and network power are the most important currencies of our age. The quants and financial engineers have invaded the big banks, and it is the venture capitalists of Palo Alto, not fund managers on Wall Street, who finance disruptive innovation. Yet the worldview of the technologist is sometimes poorly understood by those outside the tech industry. Unlike economists in the twentieth century, engineers are generally not entering politics as advisers and decision makers. They tend instead to bypass politics altogether.

In all the discussion of the political and societal problems brought about by new technology, what has been missing is an understanding of the small and anomalous group of human beings who create that technology and are constantly tweaking it, tuning it, optimizing it in response to their notions of how it ought to be better. The place where many of the most influential of those humans are educated is Stanford University, and the place they tend to congregate after they graduate is Silicon Valley. If we want to understand how and why technology is changing the world, we need to better appreciate the mindset of the engineer.

The emphasis on optimization starts early, with the introductory training one receives as an engineer or computer scientist. Students in engineering are taught to think of themselves as problem solvers, always on the lookout for better solutions. In the realm of computer science, computation is the primary tool for finding these solutions. And the idea of finding solutions as efficiently and optimally as possible is inculcated early on.

In one of the standard introductory algorithms textbooks—a book that clocks in at more than a thousand pages—an algorithm is defined as a "tool for solving a well-specified *computational problem*." The task for the human writing an algorithm is to solve a given problem with the greatest speed or by using the least possible computer memory or processing power. It's notable that the entire emphasis is on producing efficient solutions to well-specified problems. Nowhere does the text invite a reader to ask what problems are worth solving or whether there are important problems that cannot be reduced to a computational solution.

Computer science was influenced by work in decision theory and linear programming that dates back to the 1940s and 1950s. George Dantzig, a pioneer of optimization methods and professor at Stanford who retired in 1997, wrote in a 2002 retrospective that "Linear programming can be viewed as part of a great revolutionary development which has given mankind the ability to state general goals and to lay out a path of detailed decisions to take in order to 'best' achieve its goals when faced with practical situations of great complexity. Our tools for doing this are ways to formulate real-world problems in detailed mathematical terms (models), techniques for solving the models (algorithms), and engines for executing the steps of algorithms (computers and software)." He dated that effort to 1947—the year he published the celebrated Simplex algorithm for linear programming—and observed that "what seems to characterize the pre-1947 era was lack of any interest in trying to optimize."

In order to optimize, computer scientists often form mathematical abstractions of the world to create computational problems. A classic problem that nearly all computer scientists encounter at some point in their education is the "Traveling Salesman Problem" (more recently renamed the "Traveling Salesperson Problem"), or TSP. In this problem, one is given a list of cities to which the salesperson must travel, visiting every city only once before returning home. There are "costs" associated with traveling from city to city, and the point of the problem is to find a path for the salesperson that minimizes the total cost of the trip. Although to some the problem seems simple enough, it turns out that finding an efficient algorithm that can always determine the least costly trip is notoriously hard—so hard, in fact, that there is a $1 million prize for either finding an efficient algorithm to solve the problem or proving that an efficient algorithm doesn't exist. After decades of effort, the prize is still unclaimed.

Of course, many algorithms have been proposed that can find reasonably good solutions to TSP, even if they can't always guarantee an optimal result. Some of them try different alternative paths—potentially billions of them—noting when an alternative path leads to a lower cost than the best previously known path. Simpler methods take a "greedy" approach ("greedy algorithms" in computer speak) in which they simply select the next city to travel to based on the lowest cost from the current location before returning home.

Though it might seem odd that solving the traveling salesperson problem would be worthy of a $1 million prize, its real power is that it's surprisingly representative of a large class of problems, known as NP-complete problems, that underlies challenges as diverse as cryptography and DNA sequencing. Figure out an efficient algorithm to optimally solve TSP, and you will not only claim a $1 million prize but will also be able to break many encryption systems currently in use on the internet.

The choices made when creating abstractions of the world have real-world consequences. In TSP, if we choose the "costs" for traveling from city to city to be the dollar cost of airline tickets or gasoline for a car, we might make entirely different routing decisions than if the costs were based on the amount of carbon emissions produced during travel. If you've ever been puzzled why an airline travel site would suggest a trip from Los Angeles to Seattle with a layover in Chicago, it's because the objective its algorithm was set to optimize was likely to minimize price rather than environmental impact.

What we choose to optimize raises the issue of *representational adequacy*. In order to optimize some quantity (ticket cost, travel time) we need to have a way to represent that quantity mathematically. If we can't directly measure or represent that quantity, there's no way to create an optimization method to find out it if it's doing better or worse. And some things—simple things—are easier to measure than others. Measuring ticket cost is easy; measuring environmental impact is much harder. Even more difficult is determining how to optimize for more fundamental ideals such as justice, dignity, happiness, or promoting an informed democracy.

Despite these limitations, optimization is such an essential element of the computer science toolbox that it transcends simply thinking about technical problems. What begins as a professional mindset for the technologist easily becomes a more general orientation to life. It becomes second nature to perceive inefficiency, as Rob Rhinehart did with food, and get frustrated. The paramount goal becomes removing friction from everyday activities, automating repetitive tasks, and finding ways to save time while improving outcomes. There's an entire subculture of folks, many of them engineers, who traffic in "life hacking," and a popular website, Lifehacker: "the ultimate authority on optimizing every aspect of your life."

When you look at the world through this mindset, it's not just small inefficiencies that annoy. Optimization is an orientation to the big picture as well. It's not uncommon for us to encounter students, for example, who ask us for advice about how to optimize their Stanford experience, optimize their summer internship, or choose the optimal career. A recent popular book, *Algorithms to Live By: The Computer Science of Human Decisions*, recommends the skills of a computer scientist as the basis for leading a better life: algorithmic insights as a form of wisdom.

The rise of the technologist and the optimization mindset can be seen in our very lexicon. The use of the word *optimist* has been more or less constant for the past century, but a search of Google Books Ngram Viewer, which tracks the usage of particular words in a large corpus of books across languages, shows that the terms *optimize* and *optimization* were basically unknown prior to 1950 and have been on a rapid rise from then until today. This coincides with the rise of computer science as a discipline in the 1960s.

Of course, there are obvious precursors to the optimization mindset, such as the movement to bring scientific management to the workplace at the turn of the nineteenth century. This has been called Taylorism after one of its most vocal advocates, Frederick Taylor. Using empirical methods to identify best practices and standardize work in mass production lines, scientific management sought to increase worker productivity and economic efficiency. One difference between this approach and the optimization mindset of the modern technologist, however, is that a hundred years ago when bosses tried to enforce efficiency on the shop floor it was understood to be a form of oppression. Today, we tend to embrace and celebrate optimization.

But having a devotion to efficiency and an obsession with optimization is not all upside. And now that technologists have become powerful, with their vision and their values about technology remaking our individual lives and societies, the problems with optimization have become our problems, too.

THE DEFICIENCY OF EFFICIENCY

A focus on efficiency and optimization can lead technologists to believe that increasing efficiency and solving problems optimally are inherently good things. There's something tempting about this view. Given a choice between doing something efficiently or inefficiently, who would choose the slower, more wasteful, more energy-intensive path?

However, there are times when inefficiency is preferable: putting speed bumps or speed limits onto roads near schools in order to protect children; encouraging juries to take ample time to deliberate before rendering a verdict; having the media hold off on calling an election until all the polls have closed; or pursuing malicious goals—such as harming people—with greater efficiency. The quest to make something more efficient is not an inherently good thing. Everything depends on the goal or end result.

The real worry is that giving priority to optimization can lead to focusing more on the methods than on the goals in question. If a particular problem to be solved is simply delivered to a software engineer without consideration or debate as to its value, we are stuck with the results of optimizing the goal. And suddenly boosting screen time, increasing click-through rates on ads, promoting purchases of an algorithmically recommended item, increasing predictive accuracy in facial recognition, or maximizing profit leads to other important values being lost.

In our experience, the job interview questions for software engineers typically involve scalable solutions to abstract coding problems. This encourages young candidates to focus on scale and algorithmic efficiency. It does not encourage them to think critically about the company or its societal impacts when looking for a job.

The pioneering computer scientist Donald Knuth famously said that "premature optimization is the root of all evil." We could interpret this quote any number of ways, and indeed it has been widely misunderstood. Knuth uses optimization and efficiency synonymously: to optimize code is to make it more efficient. Knuth himself was not anti-efficiency; rather, he was arguing that there is a time and a place for efficiency. Usually, he observed, the biggest improvement in the running time of a program comes from modifying a small part of the code, as long as you optimize the right snippet. And making code efficient often makes it needlessly complicated, lowering overall performance because the code becomes harder to debug and maintain. An unchecked drive to make code efficient, Knuth argues, can actually *create* inefficiency for the programmer.

For Knuth, the time and place to be efficient is when you have figured out what is *worth* making efficient by analyzing the effects of efficiency at a higher level. But suppose that a technologist focuses on those bits of code that will make the code faster and writes the program in a way that doesn't sacrifice code readability. Overall, the program becomes more efficient. Has the technologist avoided premature optimization? Not necessarily. Suppose the program is used for an undesirable end: suppose it is used successfully by a malicious hacker or to guide an economy into a state of massive inequality. Perhaps then we could interpret Knuth's quote even more broadly than he intended: even if technologists know how to make a program run efficiently, if they have not considered its possible uses and impacts, optimization may yet be premature.

This points to another kind of problem that comes up when goals are not assessed. Technologists sometimes describe the tools they build as dual or multi-use. There is frequently no fixed goal or purpose for a particular technology; it can be deployed for different purposes by different users. The phrase originated in geopolitical and diplomatic circles to distinguish between civilian and military purposes of different technologies. For example, the engineers who developed nuclear energy technologies recognized that their work could be put to potentially positive uses for civilian purposes in nuclear power plants and potentially negative uses in constructing and detonating nuclear weapons. It required the scientific ingenuity of engineers to develop nuclear energy, but it was imperative that the technology be accompanied by a governance framework that would facilitate civilian uses and limit military ones.

The same is true of the digital inventions of twenty-first-century technologists. Computer scientists can create an astonishing new tool—say, facial recognition—and it can be optimized for the highest possible accuracy in identification. In a limited sense, the objective function here is to recognize a face. But having succeeded at that task, the downstream applications are many, and the ends to which the technology can be put range across a wide spectrum of good to bad. Facial recognition can be deployed so that photos can be autotagged or you can unlock your smartphone by looking into its camera. Or it can be installed on consumer drones for you to surveil your neighbors or the government to track peaceful protesters or radical terrorists to create weaponized drones that kill with newfound accuracy. What responsibility does the technologist have for trying to ensure that a technology is used for good rather than bad?

Such questions are becoming more and more urgent in tech companies as engineers increasingly grapple with the unexpected ways in which their work may be used. In 2018, thousands of employees at Google signed a letter in protest against the decision by executives at the company to sell artificial intelligence technology to the US military in order to assist in identifying people in video images. "We believe," the petitioners wrote, "that Google should not be in the business of war." In this instance, at least, Google decided not to renew its contract with the government.

It's worth emphasizing that whoever makes the choice of what to optimize is effectively deciding what problems are worth solving. The glaring lack of racial and gender diversity in the ranks of technologists and start-up founders means that these choices rest in the hands of a small group of people not representative of the wider world. No surprise that many new start-ups show a bias in favor of solving the problems of a privileged demographic. A more diverse group of technologists and founders might well deploy the power of optimization for a broader set of problems.

WHAT IS MEASURABLE IS NOT
ALWAYS MEANINGFUL

Let's assume that the technologist has successfully addressed the first problem. The focus on optimization has been accompanied by scrutiny and an independent evaluation of the goal or objective function. With some level of confidence that the problem is worth solving, the technologist gets to work.

Technologists are always on the lookout for quantifiable metrics. Measurable inputs to a model are their lifeblood, and like a social scientist, a technologist needs to identify concrete measures, or "proxies," for assessing progress. This need for quantifiable proxies produces a bias toward measuring things that are easy to quantify. But simple metrics can take us further away from the important goals we really care about, which may require complicated metrics or be extremely difficult, or perhaps impossible, to reduce to any measure. And when we have imperfect or bad proxies, we can easily fall under the illusion that we are solving for a good end without actually making genuine progress toward a worthy solution.

The problem of proxies results in technologists frequently substituting what is measurable for what is meaningful. As the saying goes, "Not everything that counts can be counted, and not everything that can be counted counts."

There is no shortage of examples of this phenomenon, but perhaps one of the most illustrative is an episode in Facebook's history. Facebook vice president of ads and business platform Andrew Bosworth revealed in an internal memo in 2016 how the company pursued growth in the number of people on the platform as the one and only relevant metric for its larger mission of giving people the power to build community and bring the world closer together. "The natural state of the world," he wrote, "is not connected. It is not unified. It is fragmented by borders, languages, and increasingly by different products. The best products don't win. The ones everyone use win." To accomplish its mission of connecting people, Facebook simplified the task to growing its ever more connected user base. As Bosworth noted, "The ugly truth is that we believe in connecting people so deeply that anything that allows us to connect more people more often is *de facto* good." He then catalogued the controversial strategies that Facebook had deployed in order to grow the user base: "the questionable contact importing practices. All the subtle language that helps people stay searchable by friends. The work we will likely have to do in China some day." He even acknowledged that connecting people would not always be beneficial: "Maybe it costs a life by exposing someone to bullies. Maybe someone dies in a terrorist attack coordinated on our tools." After the memo was published, Mark Zuckerberg repudiated it and Bosworth apologized, saying he was simply being provocative.

Economists have long worried about the problem of proxies, especially in situations where employees receive incentives for meeting their targets. In such situations, employees will quickly orient themselves not toward the worthy end but toward the proxy. The metric becomes the goal, and the means justify the end. This is called Goodhart's Law, which states that when a measure becomes a target, it ceases to be a good measure. A common sequence would look like this: The boss says we must make progress toward a large and difficult-to-measure goal. The leaders of the company choose some proxies that seem to have a plausible connection to the goal. Employees are tasked with making progress on these proxies. Though they may recognize that the given proxy is a useful but imperfect measure for the goal, they begin to lose sight of the imperfections of the proxy and thereby of the good goal, too. Soon enough, the proxy becomes the only thing that matters.

WHAT HAPPENS WHEN MULTIPLE VALUABLE GOALS COLLIDE?

Assume that the goal chosen by the technologist is worthy and that the proxies to bring about that goal more efficiently have been carefully selected and indeed effectively solve the problem. We should still ask: What happens if the engineer succeeds in optimizing one goal while ignoring other relevant values that are affected by this very success? This is the lesson of Soylent. Rob Rhinehart engineered a product that maximizes the nutritional goal of feeding oneself but neglected to consider the rival values associated with eating a meal made of actual food.

In some cases, technologies are developed to solve single and self-contained problems, so this issue never comes up. Take, for example, the heralded successes of artificial intelligence when applied to games such as checkers, chess, and Go, where machines have beaten human world champions. These are impressive technical accomplishments. And we celebrate them because the purpose of the game is clear cut: to win. Other than winning, there are no competing goals in competitive chess.

But in any circumstance involving broader aims, where our technologies interact with our personal, social, and political lives, we are typically seeking to keep multiple goals in balance at the same time. The realm of worthy ends is vast, and when it comes to world-changing technologies that have implications for fairness, privacy, personal safety, national security, justice, human autonomy, freedom of expression, and democracy, there is no guarantee that all values will happily coincide and form a grand and unified whole. It's much more realistic to assume that in many circumstances, values will conflict, and so our solutions involve a precarious trade-off among competing values.

This problem is a kind of "success disaster." The issue is not that the technologists have failed in accomplishing something but rather that their success in solving for one particular task has wide-ranging consequences for other things we care about. Think, for example, of the amazing technological advances in agriculture that have massively improved productivity. Factory farming has not only transformed the practices of growing vegetables but also made possible the widespread and relatively cheap availability of meat. Whereas it once took fifty-five days to raise a chicken before slaughter, it now takes thirty-five, and an estimated 50 billion chickens are killed every year—more than 5 million every hour of every day of the year. But the success of factory farming has generated terrible consequences for the environment (massive increases in methane gases that are contributing to climate change), our individual health (greater meat consumption is correlated with heart disease), and public health (greater likelihood of transmission of novel viruses from animals to humans that could cause a pandemic).

"Success disasters" abound in Silicon Valley technologies as well. Facebook, YouTube, and Twitter have succeeded in connecting billions of people in a social network, but now that they have created a digital civic square, they and not the government have to grapple with the conflict between freedom of expression and the spread of misinformation and hate speech. The problems created by optimizing technologists arise not because their companies have failed but rather because they have succeeded profoundly at something and become powerful as a result.

The bottom line is that technology is an amplifier. It requires us to be *explicit* about the values we want to promote and how we trade off among them, because those values are encoded in some way into the objective functions that are optimized. Technology is also an amplifier because it can often enable the execution of a particular policy to reach a goal far more efficiently than a human can. It can power an autonomous vehicle to drive more safely than your neighbor does or be the basis of a recommendation system that keeps you watching online videos far longer than you intended. Even well-meaning policies can easily become objectionable when technology enables their hyper-efficient automation. With current GPS and mapping technology it would be possible to produce vehicles that would automatically issue a speeding ticket every time the driver exceeded the speed limit—and would eventually stop the car from moving and issue a warrant for the driver's arrest when he or she had accumulated enough speeding tickets. Such a vehicle would provide extreme efficiency in upholding traffic laws related to safe driving speeds. However, this amplification of safety would infringe on the competing values of autonomy (to make our own choices about safe driving speeds and the urgency of a given trip) and privacy (not to have our driving habits constantly surveilled).

● ■ ● ■ ●

It was not that long ago that computer scientists were hackers in a garage, meeting up in computer clubs to show off their latest feats of engineering.

They were people who exercised little political power and had limited social standing. There was even a time when computer science departments struggled to attract students. But over the past thirty years, the programmer Davids have defeated the industrial Goliaths and become the new masters of the universe. Enrollments in computer science classes are booming almost everywhere. The reason is obvious: programming and data science are hugely valuable, and students want a chance to contribute to the digital revolution that is profoundly reshaping our world, changing individual human experience, social connection, community, and politics at a national and global level. Of course, the salary premium and chance of amassing start-up riches don't hurt, either. Today's lists of billionaires are topped by tech CEOs. Few people can name the CEO of an investment company; almost everyone knows the names of the founders of Microsoft, Apple, Amazon, Facebook, and Google.

Tech innovators and leaders have also become among the most powerful people in the world, not only by becoming wealthy but increasingly through their political influence as well. As their power grows, it becomes ever more important that the rest of us understand the potential problems that may result. The problems are worrisome enough when they reveal themselves in the technological innovations that have revolutionized so many aspects of our lives. They are more worrisome still when the optimization mindset is directed beyond questions of technology toward questions of social and political life. Why is that so?

Several years ago, Rob received an invitation to a small dinner. Founders, venture capitalists, researchers at a secretive tech lab, and two professors assembled in the private dining room of a four-star hotel in Silicon Valley. The host—one of the most prominent names in technology—thanked everyone for coming and reminded us of the topic we'd been invited to discuss: "What if a new state were created to maximize science and tech progress powered by commercial models—what would that run like? Utopia? Dystopia? Ultimately human evolution propellant?"

A researcher at the secretive lab leapt into the conversation. Not a hypothetical question, he said. We have already spec'd this out! Your first thought is that we should find an island and build there, but it turns out to be difficult to optimize for scientific discovery on islands. Creating infrastructure is difficult. You have to find a plot of land elsewhere, and all the desirable plots of land are already occupied. So the first problem you confront is: What do you do about the natives? We decided that the best approach is to just pay them to leave.

The conversation progressed, with enthusiasm around the table for the establishment of a small nation-state dedicated to the maximal progress of science and technology. Rob raised his hand to speak. "I'm just wondering, would this state be a democracy? What's the governance structure here?" The response was quick: "Democracy? No. To optimize for science, we need a beneficent technocrat in charge. Democracy is too slow, and it holds science back."

CHAPTER 2

THE PROBLEMATIC MARRIAGE OF HACKERS AND VENTURE CAPITALISTS

In 1996, at the World Economic Forum in Davos, Switzerland, John Perry Barlow—a lyricist for the Grateful Dead, onetime cattle rancher, and a cofounder of the Electronic Frontier Foundation—penned "A Declaration of the Independence of Cyberspace." Reacting to the passage in the United States Telecommunications Act of 1996, Barlow channeled the techno-libertarian spirit, writing "Governments of the Industrial World, you weary giants of flesh and steel, I come from Cyberspace, the new home of Mind. On behalf of the future, I ask you of the past to leave us alone. You are not welcome among us. You have no sovereignty where we gather." Taking things one step further, he espoused a utopian view of the possibilities afforded by the online world: "We are creating a world that all may enter without privilege or prejudice accorded by race, economic power, military force, or station of birth." Barlow's words were celebrated by the hackers of his time, including the young Aaron Swartz, not yet a teenager.

Few people would have anticipated the sheer concentration of power in the hands of a tiny number of private technology companies that now, more than twenty years later, determine what content we see, how we see it, and what—if any—control we have over it. In 1996, the top five companies by market capitalization were General Electric, Royal Dutch Shell, Coca-Cola, Nippon Telegraph and Telephone, and ExxonMobil; in 2020, the top five were Microsoft, Amazon, Apple, Alphabet, and Facebook. The utopian notions of technology as a great equalizer have given way to dystopian stories of data breaches, surveillance capitalism, biased algorithms, and rampant misinformation. So how did we end up here? It's a far cry from the free and decentralized internet that the early pioneers imagined.

To understand the forces at play, we need to examine the evolution of the personal computing industry from its roots in the counterculture of the 1960s to its current role as the powerhouse of the global economy. These days, the high-tech ecosystem is fueled by capital—the availability of money for investment in new companies—which creates nearly endless opportunities for building new businesses that disrupt the old economic order. In the early days of Silicon Valley, federal funding drove the development of the semiconductor industry, which laid the foundation for the development of personal computers. But federal funding was soon replaced by venture capital as a driving force in the valley's growth. Sand Hill Road, just up the hill from Stanford, is ground zero for some of the most storied names in venture capital, with billions of dollars at the ready to invest in the next world-changing idea of a twenty-two-year-old down the hill.

Though "venture capital" might evoke the stuffy world of finance, the venture industry fueling Silicon Valley has a decidedly technical mindset. Eugene Kleiner, a cofounder of the legendary venture capital firm Kleiner Perkins, was an engineer by training, who after cofounding Fairchild Semiconductor in the 1950s went on to be an early investor in Intel. John Doerr, now the chairman of Kleiner Perkins and an early investor in Amazon, Google, and Netscape, holds degrees in electrical engineering and began his career at Intel. A key figure in the formation of the dot-com bubble of the late 1990s, Doerr claimed that "we witnessed (and benefited from) the largest legal creation of wealth on the planet." Realizing that his statement had contributed to dot-com mania and a mercenary sense that making mountains of money was what mattered, he would later offer an apology.

Venture capitalists brought their own innovations to the companies they funded, and making lots of money certainly featured heavily. To more directly align the interests of employees with those of founders and investors, the inclusion of stock options in compensation packages became standard for rank-and-file engineers. As one Silicon Valley company founder and former venture capitalist told a new Stanford graduate he was recruiting, "No one ever got wealthy from salary. For that you need equity." Fresh out of graduate school in the late 1990s, Mehran saw that firsthand. Interviewing for a software engineering position at a small start-up, he finished the day by meeting with one of the company founders, a serial entrepreneur, who started the interview by saying, "I don't have any questions for you. But I can tell you that we're all going to be fucking rich." He wasn't wrong. Less than two years later, that company would hold an initial public offering (IPO) and reach a market capitalization of more than $10 billion, making all its early engineers millionaires—at least on paper for a short time. The dot-com crash of the early 2000s would begin only a few months later.

The potential for gaining riches from a future IPO adds more tinder to the fire that fuels the breakneck pace of technological development. Even established companies have used the lure of equity to grow their ranks. Microsoft recruiters, for example, would ply new candidates with projections of the potential future value of stock options.

With engineers often at the helm of companies and the firms funding them, it's no surprise that the optimization mindset would come to play an important role in how these companies are managed. In his book *Measure What Matters: How Google, Bono, and the Gates Foundation Rock the World with OKRs*, Doerr espoused the management principle of objectives and key results (OKRs), a concept originally developed by Andy Grove at Intel and now widely used at a number of technology companies, including Google, Twitter, and Uber. OKRs are the metrics that drive performance evaluations and corporate growth. As Larry Page noted in the foreword to Doerr's book, "OKRs have helped lead us to 10x growth, many times over." And an increase in a business's profitability translates into higher stock prices and greater wealth for those granted stock options.

It also translates into reward systems that shine a spotlight on engineers who help companies realize their OKRs. In 2004, Google initiated the Founders' Award to recognize teams that made substantial contributions to the company. One of the first such awards was to a team of ten engineers who had worked on the company's early ad-targeting system. The award, presented at a company all-hands meeting, led to audible gasps when the prize amount was revealed: $10 million to be split among the team.

The marriage of technology and capital has come to define the "move fast and break things" culture of Silicon Valley. The countercultural notion of a free and uncontrolled cyberspace has given way to the new mantra of "blitzscaling," where companies grow as quickly as possible to grab a dominant market position, demonstrate hockey-stick growth to their investors, and lock in any potential network effects before competitors can respond.

The monopolistic tendencies of two-sided markets that often appear on the internet only serve to reinforce the "winner-take-all" dominance of the largest players in each market. Name the largest online auction site? eBay. Name the second largest? Who knows. Buyers want to go to the place with the most sellers, and sellers want to go to the place with the most buyers. And when the buyers at those sites are advertisers, the product they're buying is you—or, more accurately, your attention. Conveniently, tech pioneers are happy to invert economic orthodoxy when it serves their interests. Peter Thiel, a founder of multiple venture capitalist firms, believes in monopoly, at least in technology markets. "Competition," he says, "is for losers."

THE ENGINEERS TAKE THE REINS

The oft-told history of Silicon Valley as an epicenter of technological innovation traces its way back to Frederick Terman, a Stanford engineering professor who after World War II would go on to serve as dean of engineering and ultimately provost. Terman encouraged both students and faculty to spend time in industry or even start their own companies. In one of the seminal stories in the creation of Silicon Valley, two of his most famous mentees, William Hewlett and David Packard, started their eponymously named company in a garage with Terman's encouragement.

The early emphasis on entrepreneurship, especially the focus on having researchers and engineers rather than business school graduates as company founders, laid a blueprint for how the engineering mindset would come to play a critical role in many subsequent tech companies. Whereas the first wave of Silicon Valley tech companies focused on hardware—semiconductors, microprocessors, and personal computers—with companies such as Fairchild, Intel, and Apple, it was only a matter of time before software—ethereal bits, not physical atoms—would become the dominant force in Silicon Valley's growth.

Fast forward to 1989, half a world away from California. The mild-mannered British scientist Tim Berners-Lee, working at the CERN laboratory in Geneva, Switzerland, proposed the creation of the World Wide Web as a means of sharing research data between labs worldwide. In his acceptance speech after winning the 2016 A. M. Turing Award—regarded as the Nobel Prize of computer science—Berners-Lee recounted that the idea for the Web had initially received little fanfare and he had been heartened that his supervisor at the time hadn't canceled the project altogether.

A confluence of events transformed a set of technical protocols developed by Berners-Lee for posting data from physics experiments into the basis for one of the greatest technological and business transformations of our lifetimes. In the early 1990s, several private internet service providers were making the Web accessible to millions and then billions of people. Many of us remember getting CD-ROMs in the mail touting America Online (later renamed AOL) or CompuServe. And by 1995, the last vestiges of government restriction on commercial use of the internet had been eliminated, making possible the "dot-com boom." Around the same time, a pair of young engineers, Marc Andreessen and Eric Bina, were busy working on the Mosaic browser at the University of Illinois Urbana-Champaign. Released in 1993, Mosaic helped bring the Web to public consciousness, transforming it from a resource for academic data sharing into something that the public could readily access. Andreessen would shortly thereafter go on to become one of the cofounders of Netscape Communications along with Jim Clark, a former Stanford engineering professor and the founder of Silicon Graphics, and release the Netscape Navigator browser in December 1994, bringing the Web to the masses. Within a year after Navigator's release, Netscape held its massively successful IPO, giving Andreessen, at the age of twenty-four, a net worth of over $50 million. A few months later he would grace the cover of *Time* magazine, sitting barefoot on a throne next to the headline "The Golden Geeks: They invent. They start companies. And the stock market has made them instantaires."

In New York during that same period, a thirty-year-old named Jeff Bezos began to realize the commercial possibilities of the internet. Bezos, who had graduated from Princeton University in 1986 with degrees in electrical engineering and computer science, had originally gone to Wall Street to use his analytical skill at quantitative hedge funds such as D. E. Shaw & Co., where he quickly rose to the position of vice president. Working through the decision as to whether he should stay at his lucrative job or make the jump to starting an internet company, he invoked a "regret minimization framework": he based his decision on minimizing the future regret he might have with regard to the decision he was making at the time. He says that the framework made his decision clear. In 1994, he left D. E. Shaw and drove cross-country to Seattle, where he started Amazon.com. The decision to found the firm was essentially the result of solving a mental optimization problem. And as they say, the rest is history. Bezos would go on to become the richest person in the world in 2018 with a personal fortune of more than $150 billion. He was in familiar company: by 2020, eight of the world's ten wealthiest people had amassed their fortunes through technology companies.

THE ECOSYSTEM OF VENTURE CAPITALISTS AND ENGINEERS

Though engineers often develop the technical breakthroughs that can serve as the seeds for the formation of world-changing companies, the growth of those seeds into blooming companies relies on access to capital, the funding necessary to hire more talent, buy equipment, secure office space, and grow the business. For many companies, venture capitalists— VCs for short—are the source of that funding. Sand Hill Road, a street roughly six miles long, is home to more than forty venture capital firms. In 1972, Kleiner Perkins Caufield & Byers—now known as just Kleiner Perkins—opened its doors as the first VC firm on Sand Hill. In 1980, it hired John Doerr away from Intel, where he had internalized the concept of OKRs. Doerr recounted the concept as articulated by Grove: "the key result has to be measurable. But at the end you can look, and say without any arguments: Did I do that or did I not do it? Yes? No? Simple. No judgments in it." OKRs can be instrumental in driving the profitability of a business by focusing on measures such as growing click-through rates on ads, increasing the amount of time users spend on a website, and attracting greater numbers of users to an app. For start-ups, showing rapid growth on such measures can be critical to attracting the capital needed to survive.

Doerr became a legend in the VC world, investing $8 million in Bezos's fledgling company in 1996. At the time, many people joked that Amazon.com should change its name to Amazon.*org* because it wasn't a profitable business and was likely to remain that way. Doerr understood its long-term potential and secured ownership of more than 10 percent of the company for Kleiner Perkins, joining the Amazon board of directors through his investment. He would later have the distinction of being one of the first VCs to invest in Google in 1999, shortly after the company's founding. At the time, both Kleiner Perkins and Sequoia Capital—the other eight-hundred-pound gorilla on Sand Hill—were vying hard to be the sole VCs to invest in Google. Larry Page, the cofounder and twenty-six-year-old dropout from Stanford's computer science PhD program, gave them an ultimatum: they could either invest together or walk away from the deal. They became coinvestors. Doerr plunked down $11.8 million—his "biggest bet in nineteen years as a venture capitalist"— to get 12 percent of the company and a seat on the board.

Shortly thereafter, Doerr arrived at the Google offices with what he described as a "present," the introduction of OKRs as a management tool. As Doerr told the story, he came with a presentation in which his "first PowerPoint slide defined OKRs: 'A management methodology that helps to ensure the company focuses efforts on the same important issues throughout the organization.'" He went on to explain that an "*OBJECTIVE* . . . is simply *WHAT* is to be achieved, no more and no less. . . . *KEY RESULTS* benchmark and monitor *HOW* we get to the objective. . . . Most of all, they are measurable and verifiable."

The executive staff at Google came to embrace the idea wholeheartedly. It is, after all, an engineering approach to management. Even paralleling the language of optimization theory, in which an objective is a function whose value is to be optimized, the idea of having measurable outcomes on which to focus a company's efforts fit the engineering-oriented management team like a glove.

The use of OKRs became the standard by which nearly everyone at Google was measured. As Doerr wrote, "the marriage of Google and OKRs was anything but random. It was a great impedance match, a seamless gene transcription into Google's messenger RNA. OKRs were an elastic, data-driven apparatus for a freewheeling, data-worshipping enterprise." Each quarter, engineers, salespeople, researchers, and product managers would measure how well they had succeeded against their existing OKRs and then draw up plans for the OKRs they would focus on for the coming quarter. All-hands meetings would report on how the organization had fared on its company-level OKRs. They provided a concrete, measurable, and objective metric for how individuals and the company were doing.

Doerr is an evangelist for OKRs—and not only at Google. In his book *Measure What Matters*, he recounted how OKRs have helped numerous companies and nonprofits create more transparency in their culture, determine when course corrections are necessary, and achieve audacious stretch goals. Indeed, Larry Page credits Google's massive growth in part to OKRs, writing in the foreword to Doerr's book that "I think it's worked out pretty well for us."

Of course, OKRs are not the only management philosophy among big tech companies, although they are espoused by many of them. But they are emblematic of how readily the engineering mindset of measurement and optimization has ballooned beyond the solution of technical problems. Rather, as engineers have assumed the role of company leaders and have become venture capitalists themselves, the engineering mindset has moved to the highest levels of corporate governance. Examining the impact of that mindset is critical to understanding how questions of human well-being and societal flourishing may—or may not—be taken into consideration in the decision-making processes of tech companies.

THE OPTIMIZATION MINDSET MEETS CORPORATE GROWTH

Though management tools such as OKRs coupled with an optimization mindset have fueled enormous corporate growth and led to the creation of billions of dollars in shareholder value, they also raise important questions: How are the objectives to measure to be chosen? What business and technical choices must be made in the drive to optimize them? And how far should those decisions be taken?

The use of OKRs in Google's YouTube subsidiary was explained by Vice President of Engineering Cristos Goodrow while recounting a realization he had in 2011:

> As Microsoft CEO Satya Nadella has pointed out: In a world where computing power is nearly limitless, "the true scarce commodity is increasingly human attention." When users spend more of their valuable time watching YouTube videos, they must perforce be happier with those videos. It's a virtuous circle: More satisfied viewership

(watch time) begets more advertising, which incentivizes more content creators, which draws more viewership.

Our true currency wasn't views or clicks—it was watch time. The logic was undeniable. YouTube needed a new core metric.

To argue for this new metric, he wrote an email to the YouTube executive team arguing that "Watch time, and only watch time" should be the objective to improve at YouTube. In essence, he equated watch time with user happiness: if a person spends hours a day watching videos on YouTube, it must reveal a preference for engaging in that activity. But the simple fact that we engage in activities is not necessarily an indicator that these activities—including things such as doing dishes, mowing the lawn, or even smoking—make us happy or contribute to our well-being. Yet the focus on watch time ultimately became the basis of one of YouTube's most significant objectives: to reach 1 billion hours of watch time per day by 2016—a goal it ultimately surpassed.

To be fair, Goodrow notes that in pursuing its goal, YouTube did sometimes take actions that had a negative impact on watch time if the company believed that the action was in the user's interest: "For example, we made it a policy to stop recommending click-baity videos." But he followed up by saying "We never did *anything* without measuring impact on watch time." Left out seem to be questions such as whether it's really healthy for children (or adults, for that matter) to watch an endless stream of videos; whether conspiracy theory videos by flat-earthers should be recommended with the same gusto as more benign videos; or what the race for watch time could do to the ecosystem of content producers—who are paid by advertisers when their videos are watched—who might create more outrageous videos in order to have their content be the centerpiece of the user's coveted watch time.

Doerr himself acknowledged that management systems such as OKRs can have their faults, writing, "Like any management system, OKRs may be executed well or badly." He even affixed a warning label in his book, originally suggested by a Harvard Business School paper with the cheeky title "Goals Gone Wild": "Goals may cause systematic problems in organizations due to narrow focus, unethical behavior, increased risk taking, decreased cooperation, and decreased motivation. Use care when applying goals in your organization." What is missing is an explicit call for moral commitments to be an objective in themselves, rather than just a side consideration in the pursuit of some more directly measurable goal. The fact that it can be impossible to clearly measure the "social goodness" of an outcome means that it becomes difficult to factor it into the metric being optimized. It's easier to just assume that watching more videos must be making users happier. Measuring watch time is straightforward; determining whether users are actually happier, more factually informed, or politically radicalized, is not.

OKRs are just a more recent manifestation of the optimization mindset in business. Examples abound of how a rush to focus on maximizing a particular objective can lead to worse outcomes for things we might care about. One of the earliest computer manufacturers, Digital Equipment Corporation, wanting to improve its customer service, installed a system to monitor the average time it took its customer service representatives to respond to help center calls. The average call time was displayed for its customer service representatives to see. When the time grew too long, the representatives simply answered the phone, saying "Our systems are down right now, please call back later." That led to a rapid decrease in the average call time, which was being measured, and an increase in customer irritation, which was not.

As Lisa Ordóñez and her colleagues explained in their article "Goals Gone Wild: The Systematic Side Effects of Overprescribing Goal Setting," overreliance on goal setting can cause individuals and organizations to become so focused on narrow goals that they lose sight of other important considerations that need to be balanced when building products. This myopic view can in turn lead to all manner of bad outcomes, including excessive risk taking and an increase in unethical behavior, in order to meet the demands of a goal. In the longer term, a narrow focus on goals can erode the culture of an organization, as broader interests give way to simply meeting specific quantitative targets. The researchers' work is replete with examples. Perhaps the most famous is the tragedy of the Ford Pinto. To quote Ordóñez and her colleagues:

> CEO Lee Iacocca announced the specific, challenging goal of producing a new car that would be "under 2,000 pounds and under $2,000" and would be available for purchase in 1970. This goal, coupled with a tight deadline, meant that many levels of management signed off on unperformed safety checks to expedite the development of the car—the Ford Pinto. One omitted safety check concerned the fuel tank, which was located behind the rear axle in less than 10 inches of crush space. Lawsuits later revealed what Ford should have corrected in its design process: The Pinto could ignite upon impact. Investigations revealed that after Ford finally discovered the hazard, executives remained committed to their goal and instead of repairing the faulty design, calculated that the costs of lawsuits associated with Pinto fires (which involved 53 deaths and many injuries) would be less than the cost of fixing the design. In this case, the specific, challenging goals were met (speed to market, fuel efficiency, and cost) at the expense of other important features that were not specified (safety, ethical behavior, and company reputation).

Had Ordóñez written her article a decade later, she could easily have included more examples from the tech world, such as whether an extreme focus on increasing video watch time may have left little room for considering the political, social, and health impacts of millions of people being glued to their screens. Of course organizations need to establish goals. But a relentless focus on trying to optimize what's quantifiable within the narrow view of any organization fixated on growth doesn't necessarily provide insight about what's good for individuals, society, or the world.

The results of the myopic focus in designing the Ford Pinto were deadly for dozens of people. The impact of fixating on a poorly chosen objective or misleading metric in today's tech-driven world, though often not deadly, has more far-reaching social consequences. When click-through rates increase on untruthful ads that question the integrity of election results without evidence or promote conspiracy theories about vaccines, revenues and market capitalizations may rise, but democracy and the well-being of hundreds of millions of people will suffer. Even when there is fallout for the companies responsible, it's often short lived. Consider that Facebook's stock price did not suffer any long-term damage from the Cambridge Analytica scandal in 2016. To the contrary, it has since soared. If the market rewards only revenue, where's the incentive to protect democracy or other values we cherish?

HUNTING FOR UNICORNS

Just over fifty years ago, Milton Friedman—still six years shy of winning the Nobel Prize in Economics—penned an essay in the *New York Times* espousing the view that "The Social Responsibility of Business Is to Increase Its Profits." His argument followed the basic premise that a business is beholden to its owners—shareholders in the case of public companies—and as a result should be trying only to maximize its value. He explicitly rejected the role of social responsibility in corporate governance, in part because of the difficulty of generating a quantifiable assessment of such responsibilities. Rather, he argued for a singular focus on return to stockholders, stating that taking into account social responsibility is tantamount to spending stockholders' money for a vague, ill-defined social interest. He concluded with an oft-quoted refrain from his earlier book *Capitalism and Freedom*: "There is one and only one social responsibility of business—to use its resources and engage in activities designed to increase its profits so long as it stays within the rules of the game, which is to say, engages in open and free competition without deception or fraud." To Friedman, it shouldn't matter to a company whether its customers spend all their time scrolling through misinformation on a social network or watching endless online videos. As long as the company is legally promoting that behavior to maximize return to its investors, it is doing the right thing. Perhaps the sociologist C. Wright Mills characterized it most starkly well before Friedman's influential article: "For of all the possible values of human society, one and one only is the truly sovereign, truly universal, truly sound, truly and completely acceptable goal of man in America. That goal is money." Of course, Mills was describing the role of money in American life, not endorsing it.

Though the views of some people in the VC world are perhaps not as extreme as Friedman's, there is no question that VC funds are investment vehicles for their limited partners (LPs), the people or organizations who provide the capital for those funds. As a result, VCs have a fiduciary duty to provide a return on investment to them, making a drive for profit a primary motivation.

The way returns are often generated in the VC world is a curious beast. As Peter Thiel and Blake Masters noted in their book *Zero to One: Notes on Startups, or How to Build the Future,* "The biggest secret in venture capital is that the best investment in a successful fund equals or outperforms the entire rest of the fund combined." In other words, most of the profit a fund makes comes from a single company—a Google, Facebook, or Uber—whose market capitalization dwarfs that of every other company in the fund.

Of course, VCs have their own lingo for the phenomenon. The goal is to drive outsize returns by discovering future "unicorns"—companies with at least a billion-dollar market capitalization. It's an elusive task. An analysis of start-up valuations found that "it remains pretty rare for a company to hit the unicorn mark. In fact, the number looks to be just above 1% of companies." Or, according to the analysis, 1.28 percent, to be exact. Finding such deals creates competition among VC firms to seek out and build relationships early with would-be entrepreneurs. It sometimes also leads to unexpected intrusions at lunch, as Mehran found out shortly before finishing graduate school. Out with a fellow graduate student for dim sum at a Chinese restaurant not far from Sand Hill Road, they were just finishing their meal when they were approached by a man in a suit whom neither of them recognized. He politely introduced himself and said, "I couldn't help but overhear that the two of you were talking about search engines. If you decide to start a company, give me a call." Then he slid his business card across the table.

In the last decades of the twentieth century and the early part of the twenty-first, VCs, many based in the San Francisco Bay Area, assumed an outsize role in funding start-ups, especially in biotechnology and information technology. Earlier models for building companies that had relied on grants from the government or bootstrapping businesses without taking much external investment quickly gave way to an influx of readily available venture funding. That funding came with the expectation of rapid measurable growth. The number of VC firms increased by a factor of ten between the 1980s and 2000. Stanford itself got into the game, claiming early equity positions in more than eighty companies started on campus by its students or faculty, including, of course, those created by student dropouts, such as Google. By 2000, the total amount of venture fund investment exceeded $100 billion and VC-funded companies made up 20 percent of publicly traded firms in the United States and nearly a third of total US market capitalization. Although most companies that receive VC investments fail, one analysis noted that of all publicly traded companies founded after 1979, 43 percent have been VC backed and account for nearly 60 percent of total market capitalization.

The rarity of finding or helping develop a unicorn creates a strong incentive to push companies to quickly become leaders in their sectors. Reid Hoffman, the cofounder of LinkedIn and also a partner at the venture capital firm Greylock Partners, explained this through the concept of *blitzscaling*: "To prioritize speed, you might invest less in security, write code that isn't scalable, and wait for things to start breaking before you build QA tools and processes. It's true that all these decisions will lead to problems later on, but you might not have a later on if you take too long to build the product." Hoffman is no disciple of Milton Friedman, however, as he also wrote, "We believe that the responsibilities of a blitzscaler go beyond simply maximizing shareholder value while obeying the law; you are also responsible for how the actions of your business impact the larger society." But it can be difficult to fully appreciate the downstream impact of a business if there's a relentless push to get a product out the door and generate revenue. Jack Dorsey acknowledged as much, tweeting (naturally) in 2018, "Recently we were asked a simple question: could we measure the 'health' of conversation on Twitter? This felt immediately tangible as it spoke to understanding a holistic system rather than just the problematic parts. If you want to improve something, you have to be able to measure it." He went on to suggest that Twitter would try to take on that challenge. But his message came two years after Russian operatives had flooded the platform with disinformation during the 2016 US presidential election, paving the way for domestic disinformation campaigns in 2020.

In one of our classroom conversations at Stanford, Nicole Wong, who had served as vice president and deputy general counsel at Google before joining Twitter as legal director of products and eventually being appointed deputy chief technology officer of the United States in the Obama administration, discussed the emphasis on user engagement in online platforms such as YouTube. She argued for a "slow food" movement in technology in which, rather than valuing speed and engagement, platforms should seek to promote authenticity, accuracy, and context of content. In recounting her time at Google, she explained that the problems on platforms such as YouTube had become clear only once they scaled up, but perhaps slowing down the process of pushing products out the door might lead to more reflection about their potential impact and the ability to rethink the criteria that might actually indicate success.

Though the notion of taking a slower and more reflective approach toward the development of technology products is certainly worthy of consideration, it stands in contrast to the expectations of what VCs want to see from their companies: revenues and returns. "Investors are a simple-state machine," Michael Siebel, the CEO of Y Combinator, recounted to MIT Technology Review. "They have simple motivations, and it's very clear the kind of companies they want to see." A focus on "getting big fast" to reach an "exit"—either an IPO or a high-priced acquisition by an even larger company—is one of the primary goals that VC firms prioritize for their portfolio companies. In fact, many VCs use the tally of their high-value "exits" as their own measure of success. And with billions of dollars available for investment—the amount of money managed by VC firms grew from roughly $170 billion in 2005 to $444 billion in 2019—it's no wonder that the emphasis on speed of execution, metrics to measure results, and ultimately returns for investors leads to a cycle where "moving fast and breaking things" can be adopted with little reflection until it's too late.

It's also worth noting that while funding in the VC world continues to grow, how that money is distributed often reflects a narrow view of what a successful entrepreneur looks like. John Doerr, speaking at the National Venture Capital Association in 2008, famously described how being a male nerd "correlates more with any other success factor that I've seen in the world's greatest entrepreneurs. If you look at [Amazon founder Jeff] Bezos, or [Netscape founder Marc] Andreessen, [Yahoo cofounder] David Filo, the founders of Google, they all seem to be white, male, nerds who've dropped out of Harvard or Stanford and they absolutely have no social life." But looking for patterns like this in who to fund also leads to inequities in how entrepreneurs might be evaluated by the VCs they are pitching to.

Indeed, the data reveals stark inequalities in funding by founders' gender and race. CrunchBase, a well-known provider of information about start-up companies that produces an annual analysis of the distribution of venture funding, reported that only 2.3 percent of venture funding went to female-led start-up companies in 2020, a decline from 2.8 percent the previous year. To put that in context, another Crunch-Base report touted that "20 percent of global startups raising their first funding round in 2019 [have] a female founder." Similar disparities can be seen by race. From 2015 to 2020 only 2.4 percent of total venture capital funding went to Black and Latinx founders, despite the fact that data from the US Census Bureau in 2019 reports that 18.5 percent of the US population was Hispanic or Latinx and 13.4 percent of the population was Black or African American. The uneven distribution of venture funding creates inequities for who gets funded to build technologies and, in turn, who chooses what should be optimized, and for whom.

THE NEW GENERATION OF
VENTURE CAPITALISTS

In the tech world, a successful engineer is also a potential future venture capitalist, and this dual identity has expanded significantly in recent years. At Stanford, VCs showcase their new companies on the lawn across from the Gates Computer Science Building while handing out boba tea and beanies to recruit the students spilling out of classes, and engineering students pitch VCs in student-run business plan competitions. Despite offering funding to college students to drop out of college in order to start companies, Peter Thiel, like other VCs, has frequently taught classes on campus. The porous boundary between technical talent and the firms that fund them has created an unparalleled engine for growth. According to a study released by Stanford in 2011, if the companies started by Stanford graduates, frequently funded by other Stanford graduates, were their own independent nation, it would be the tenth largest economy in the world. It also creates an inward-looking and self-reinforcing system whose values become more detached from everyone outside it.

For many engineers who found themselves wealthy after the dot-com boom of the late 1990s, the next step was to become angel investors or venture capitalists themselves. Similar to the way the hardware engineers of the 1970s and '80s had brought about a shift in venture funding from Wall Street and the East Coast world of finance to Sand Hill Road and the West Coast world of software, many of the engineers who had ridden the first dot-com wave became poised to fund the next round of tech companies. Marc Andreessen would swap the jeans he had worn on the cover of *Time* for a sports jacket, founding the venture capital firm Andreessen Horowitz with his longtime colleague Ben Horowitz in 2009. Their firm would become an investor in Twitter, Instagram, Facebook, Pinterest, Lyft, and Airbnb.

In an oft-quoted 2011 piece in the *Wall Street Journal*, "Why Software Is Eating the World," Andreessen explained how the capital needs of tech companies had changed:

On the back end, software programming tools and Internet-based services make it easy to launch new global software-powered start-ups in many industries—without the need to invest in new infrastructure and train new employees. In 2000, when my partner Ben Horowitz was CEO of the first cloud computing company, Loudcloud, the cost of a customer running a basic Internet application was approximately $150,000 a month. Running that same application today in Amazon's cloud costs about $1,500 a month.

Thus the drastically reduced capital needs for software companies enabled the optimization mindset to create whole new models for venture creation.

A key idea for applying optimization to complex problems is the notion of optimizing from multiple starting points. Translated into the world of start-ups, that means that if you fund only a small number of companies, your chance of hitting on an eventual unicorn is limited. But if you fund a large number of companies, you have many more opportunities to hit a home run. Each investment gives you a new starting point for your revenue optimization process. This notion is further supported by the observation of one venture capitalist, Dave McClure, that "most investments fail, a few work out ok, and a very tiny few succeed beyond our wildest dreams." His conclusion: "If unicorns happen only 1–2% of the time, it logically follows that portfolio size should include a minimum of 50–100+ companies in order to have a reasonable shot at capturing these elusive and mythical creatures."

One of the better-known venture firms to realize the potential of vastly increasing the number of start-up investments is Y Combinator, founded in 2005. The name of the firm comes from the theory of computation and refers to a function that generates other functions. Indeed, Y Combinator's entire goal is to create other companies. Given the techie name, it's perhaps not surprising that three of the firm's four founders hold PhDs in computer science. They made their initial fortune through founding and selling a prior company, Viaweb, to Yahoo! in 1998 for $50 million.

Y Combinator—YC for short—is often referred to as a start-up "accelerator" as it not only invests in very young companies but helps to bring together small groups of entrepreneurs in "batches" to create such ventures and mentors them through the process of securing additional investment. Its standard deal is to invest $125,000 in return for 7 percent of a company, leveraging the fact that infrastructure costs for internet companies have plummeted in recent years.

YC fosters a "lean start-up" mentality, where there is a push to build a minimal viable product, get it out to potential users to find out what resonates, and then iterate quickly to try new possibilities if early ideas fail to get traction. In essence, it's applying an optimization process to find product ideas and features that consumers would consider using. Showing some early user adoption—the coveted potential for "product-market fit"—in turn maximizes the possibility of building something that will excite potential investors to take a bet on the company and provide the capital to help take the initial product to the next level and (with even more funding) to the one after that.

Nineteen-year-old Aaron Swartz was a member of YC's first entrepreneur batch in 2005. He dropped out of Stanford to continue pursuing Infogami, the company he had formed as a result of participating in the program. Unable to secure sufficient funding to continue Infogami as a stand-alone venture, he was convinced by YC executives to merge his fledgling company with another start-up, spawning Reddit and landing Swartz with the moniker "Reddit cofounder" in the process.

As with most venture capital firms, YC takes active steps to promote the success of its portfolio companies. Twice a year it holds "Demo Day," when entrepreneurs in its program present the start-ups they have been working on during the past few months to a roster of venture capitalists and angel investors—often engineers who, after making lucrative "exits" from prior companies, now have substantial personal bank accounts with which to make investments in new ventures. In recent years, each Demo Day has included more than a hundred start-ups presenting to approximately a thousand attendees, many of whom apply to attend this invitation-only event. The results for the companies presenting are substantial. YC reported that "Each batch of YC companies raises about $250M of seed capital in the weeks following Demo Day." At two batches a year, that's half a billion dollars of capital a year pouring into start-ups from one organization alone. As YC recently noted on its website, "Since 2005, Y Combinator has funded over 2,000 startups. Our companies have a combined valuation of over $100B," and include names such as the gig economy companies DoorDash, Instacart, and Airbnb, as well as the self-driving car company Cruise. The YC program is so competitive that just being accepted into it is often touted as a badge of success by would-be entrepreneurs, even if their start-up ideas turn out to be failures. Not wanting to miss out on the action, in 2011 Andreessen Horowitz created a separate fund to invest $50,000 in each start-up accepted into YC's program. Evidently, having multiple starting points to optimize from can lead to handsome returns.

Despite what the tech executives of a few large companies may claim to lawmakers on Capitol Hill about their ability to balance profits while achieving other societal values, the burgeoning landscape in technology companies calls for broader controls to make sure that all players adhere to the values society wants to promote. Today, the focus may be on calling Mark Zuckerberg to testify before Congress or even launching antitrust actions against the biggest companies, but there are literally hundreds of would-be Zuckerbergs who are in the pipeline to build the next world-changing, disruptive, optimizing, and potentially socially deleterious product. It's not just a question of identifying the problems with celebrated tech founders. It's not even a question of identifying the problematic companies whose harmful social consequences need to be reined in. We need to become clear about the values we care about to understand how to set the rules for far more companies tomorrow.

TECHNOLOGY COMPANIES TURN MARKET POWER INTO POLITICAL POWER

Technology companies are now active in turning their market capital into political capital. Responding to growing calls for regulation, companies are pushing back through lobbying, public relations efforts to sway public opinion, and direct engagement with lawmakers to influence legislation. Not only have the engineers become the financiers, they are now trying to set the rules for how they are, or are not, regulated. It's an incredible shift from the early days of counterculture hackers—people who might have chatted online with John Perry Barlow about the Grateful Dead at the Whole Earth 'Lectronic Link (WELL)—to those who now look to influence the political arena for the financial benefit of their companies.

In 2008, the Illinois General Assembly passed the Biometric Information Privacy Act (BIPA), a pathbreaking piece of legislation that limits the collection and usage of biometric data, such as fingerprints and facial geometry (which can be inferred from photographs of individuals). The law requires that companies gathering such biometric data obtain written consent from their users. And the penalties for not doing so are steep: fines range from $1,000 to $5,000 *per person* for violations.

In 2015, Facebook was sued under this law based on its use of facial recognition technology to identify users' faces in photos. While fighting the case in court, where the potential for billions of dollars in penalties loomed, Facebook was also accused of trying to run an end game around the legislation itself. State Senator Terry Link, who had introduced the original BIPA legislation, proposed an amendment that would have removed information derived from photographs from the types of data that BIPA required consent for. The amendment would have eliminated the grounds for the lawsuit against Facebook. But facing sharp pushback from privacy groups and even the state attorney general, the amendment was ultimately withdrawn. The lawyers for the plaintiffs in the case alleged that "Facebook and a variety of other Silicon Valley companies have lobbied for this change." Facebook denied the lobbying claim, but public records show that it had been a contributor to several of the amendment's backers.

Ultimately, the lawsuit went forward, and by early 2020 Facebook agreed to settle to the tune of $550 million. That may seem like a remarkable sum, but it's actually quite a discount when compared to the maximum $47 billion penalty it could have faced. The judge presiding over the case felt similarly, reportedly asking, "It's $550 million. That's a lot. But the question is, is it really a lot?" Responding to the judge's concerns, Facebook increased the settlement amount to $650 million and changed its facial recognition setting to an opt-in default on systems worldwide. In February 2021, the settlement was finally approved. Of course, the judge was right in asking whether the fine was really a lot. Facebook's revenue in just the first three months of 2020 was more than $17 billion. It could easily pay the fine and move on to business as usual, not to mention continue to lobby and make campaign contributions.

The market power of large players can translate into significant influence and leeway in the political sphere. As companies are threatened with the passage of new regulations, they learn to use their influence and money to resist regulation and shape policies. As Alvaro Bedoya, the executive director of the Center on Privacy and Technology at Georgetown University, told a reporter during the BIPA case, Facebook's approach has been "If you sue us, it doesn't apply to us; if you say it does apply to us, we'll try to change the law." It's no wonder that in September 2019, Zuckerberg flew to Washington, DC, to have closed-door meetings with lawmakers. A report by *Axios* quoted a Facebook official as saying, "Mark will be in Washington, D.C., to meet with policymakers and talk about future internet regulation. There are no public events planned." Private meetings such as this, which exclude public discourse and have little oversight, allow company executives to press lawmakers for policies that are favorable to their businesses or argue that their companies are capable of "self-regulating," making new regulations unnecessary.

It's rational management to allocate your resources in a way that maximizes the impact on your business's bottom line, and that can include political power through lobbying. In 2019 and 2020, Facebook and Amazon spent more on federal lobbying than any other company, besting even defense contractors such as Lockheed Martin. As Representative David N. Cicilline of Rhode Island explained, "These companies, because they are so large, have tremendous economic power and tremendous political power. And they're spending hundreds of millions of dollars to try to protect the status quo." Big tech is also spending millions to lobby

European regulators to ward off efforts to limit digital advertising, contributing to what some call a "Washingtonization of Brussels." Those lobbying efforts are unlikely to abate anytime soon, even as those companies come under greater antitrust scrutiny.

One of the recent battlefronts in tech companies' push to influence regulation comes from California's effort to reclassify gig economy workers, such as Uber and Lyft drivers and delivery people for companies such as DoorDash, as employees rather than contractors of the firms they work for. In 2019, the California legislature passed Assembly Bill 5 (AB 5), with the aim of reclassifying thousands of independent contractors as employees, thereby guaranteeing them numerous benefits such as minimum wage, unemployment insurance, and sick leave. The bill was a textbook case of government seeking to contain a negative externality created by a profit-seeking company. Assemblywoman Lorena Gonzalez, an author of the bill, described her motivation: "As lawmakers, we will not in good conscience allow free-riding businesses to continue to pass their own business costs onto taxpayers and workers." Providing such benefits would cost the likes of Uber and Lyft millions of dollars. Their response was swift. First, they tried to get an injunction against the new law, delaying its effective date. They were denied. Then they threatened to shut down their operations in the state. In the meantime, they worked on producing a ballot initiative, Proposition 22, that defined "app-based transportation (rideshare) and delivery drivers as independent contractors and [the adoption of] labor and wage policies specific to app-based drivers and companies," effectively exempting them from the requirements of AB 5. A consortium of tech companies, including Uber, Lyft, and DoorDash, raised more than $200 million to get Prop 22 onto the ballot and sing its praises to the electorate, including pushing smartphone alerts to millions of their users, encouraging them to vote for the proposition. The proposition's text also includes a provision that prevents the legislature from amending the provisions unless it can achieve a seven-eighths majority, a nearly impossible task.

On November 3, 2020, the citizens of California voted in favor of Prop 22 by a nearly 20-point margin. The vote not only gutted AB 5 in

California but also sent a clear message to other states that attempts to provide more rights for gig economy workers would be met with well-funded lobbying campaigns and would be likely to end in failure. Though it's difficult to know the full range of reasons why voters approved the measure, the heavily funded advertising efforts of the companies whose interests were on the line were undeniable. And their ability to contact not only the drivers and delivery people who worked for them but also the users of their apps gave them a direct means of communication—and influence—with the people who would be most directly impacted by the passage of the proposition. It was effectively free political advertising anytime a user opened the app on their phone. As New York University professor Arun Sundararajan noted to the *Verge*, "I doubt whether the average voter would have weighed the pros and cons of the labor law around AB5 versus the new initiative. They feel positively towards the [tech] platforms, they don't want to see a disruption in something that they depend on, and so they vote for the platform's position." Indeed, with the claim by Lyft that "Up to 90% of app-based driving jobs could disappear" if Prop 22 did not pass, it's likely that both its drivers and the far larger number of people who depend on it for rides would vote in favor of the proposition without necessarily reading through the details of the legislation, much less taking time to weigh its implications.

The landscape of legislation and ballot initiatives impacting tech companies will only continue to grow. Battles over net neutrality, about which many consumers share the position taken by Netflix, Google, Facebook, and Amazon that internet service providers should not be able to charge differential fees for access to particular services, have now given way to the federal government's considering taking up antitrust actions against many of those same companies, often with the blessing of consumers. As Speaker of the House Nancy Pelosi put it, "Unwarranted, concentrated economic power in the hands of a few is dangerous to democracy—especially when digital platforms control content. The era of self-regulation is over." That remains to be seen. Big tech will not submit to regulation without a fight. One outcome is clear: we are in a different land from the one in which John Perry Barlow hoped we might end up.

CHAPTER 3

THE WINNER–TAKE–ALL RACE BETWEEN DISRUPTION AND DEMOCRACY

A few years ago, at another dinner in Silicon Valley, Reid Hoffman spoke freely as he offered his perspective on the growing public outrage at tech companies. If you are the CEO of a tech company, he said, your primary concern is your competitors. The tech industry's big five—Google, Facebook, Apple, Microsoft, and Amazon—are in an arms race with one another for talent, and each believes that it could be rendered extinct by the others in the blink of an eye. When Facebook moved to a new corporate campus, Mark Zuckerberg left the sign of the previous occupant, Sun Microsystems, at the entrance as a reminder to employees of how quickly a dominant company can be left behind.

Even if the market position of a company seems secure, China's big tech companies are growing like mad. *Only the Paranoid Survive* is the title of former Intel CEO Andy Grove's book for good reason. In a world marked by this profound sense of insecurity, Hoffman suggested, the threat of government regulation is an afterthought. Any CEO's goal, he explained, is to be constantly innovating.

Although Hoffman believes in the important role that government can play, a surprising number of his compatriots in the tech sector do not. The marriage of the optimization mindset and the profit motive often leads to a libertarian approach to politics and the role of government in the marketplace. Unbridled innovation can make it difficult for governments, focused on writing rules to address the needs and priorities of their citizens, to keep up. In his typically contrarian way, Peter Thiel has said, "One of the things that's striking about talking to people who are politically working in D.C. is, it's so hard to tell what any of them actually do." He argues that government is actually to blame for a deceleration in the rate of technological innovation—and that's a danger he thinks we need to avoid.

This isn't just a stereotype about Silicon Valley. The libertarian strain could be seen in John Perry Barlow's 1996 "Declaration of the Independence of Cyberspace" and in the counterculture roots of many computing enthusiasts in the 1970s and '80s. Nowadays, though, it's widespread. A recent Stanford study systematically mapped the libertarian attitudes of tech leaders—a distinctive mix of views that are socially progressive and economically conservative—who turned out to be even more hostile to regulation than were run-of-the-mill millionaires. The researchers concluded that this embrace of libertarian views manifests itself early, even among undergraduate computer science majors.

Maybe this disregard for interventionist government is well deserved. After all, politicians and policy makers often lack the expertise in emerging technologies that is required to make smart regulatory choices. Bill Foster, a physicist and one of the few members of Congress with a PhD, noted in 2012 that only about 4 percent of members in Congress had technical backgrounds. "Most members of Congress don't know enough about science and technology to know what questions to ask," said former congressman Rush Holt of New Jersey, "and so they don't know what answers they're missing." Moreover, the potential effects of new technologies are so uncertain that it's reasonable to ask whether regulators should simply react to developments rather than proactively constrain technological change. Finally, if those arguments aren't sufficient to undermine the case for an active government role, leaders in tech are prone to say something like the following: "If the US government gets in the way, then . . . China." The bogeyman of Chinese dominance in technology is enough to cow many proponents of more active regulatory oversight into submission.

Where does this antigovernment view leave the rest of us? Because if we take a step back, regulation is just a loaded word for an important thing: the actions taken by those we elect to transform our shared values (and reconcile our differences) into rules that serve the common interest. So when engineers and venture capitalists bemoan government regulation, they are in effect rejecting the role that democratic institutions serve in establishing rules of fair play, facilitating cooperation that advantages everyone, and helping us address the (potential) negative impacts of new technologies.

If democratic politics doesn't have a role to play, what is the technologists' preferred alternative? Perhaps Mark Zuckerberg's obsession with Roman emperors (he named his two daughters Maxima and August and spent part of his honeymoon in Rome taking photos of sculptures of the emperor Augustus) offers us a clue. As with Plato, there is an almost ahistorical faith in a new generation of philosopher kings—this time, a technocracy governed not by philosophers but by technologists who are motivated by the right things, have pure intentions, and can engineer amazing social outcomes if only everyone else would get out of the way.

The question is whether this is a form of governance we are willing to accept. If not, are we prepared to trade off other things, such as unfettered technological development, to ensure democratic oversight of such innovations?

INNOVATION VERSUS REGULATION IS NOTHING NEW

On March 25, 1911, in New York City, a garment factory owned by the Triangle Waist Company caught fire, killing 146 workers, one of the deadliest industrial accidents in US history. As described in lurid detail by newspapers at the time, the factory's hazardous conditions, including locked doors and windows and inadequate escape ladders, had trapped workers inside the factory as it burned. The victims, 123 of whom were women and girls, succumbed to the fire or smoke inhalation or jumped to their deaths from the windows of the ten-story building. For labor activists such as Frances Perkins, who later became the first woman appointed to a cabinet position when she served as secretary of labor under President Franklin D. Roosevelt, the fire was proof that "something must be done." More than 100,000 people attended the funeral march for the victims—a procession that marked a critical turning point in the campaign for improvements in workers' rights.

Up until that point, efforts to improve workplace safety had made little progress. Europe and North America were in the midst of an industrial revolution—with the emergence of organized machine production of various forms—and the Dickensian sweatshops of the garment industry were a leading example. The new machines were more efficient than human labor alone, and when the machines could be operated by low-paid immigrants, factory owners saw massive gains in productivity and profit. Between 1870 and 1900, employment in sweatshops doubled, capital investment tripled, and New York came to dominate the industry, producing more than 40 percent of ready-to-wear clothes in the country. Though the period also saw the founding of the International Ladies' Garment Workers' Union (ILGWU) in 1900, protests and strikes were episodic affairs that rarely captured much public attention. As the factories were staffed by young immigrant women in desperate need of employment, the concerns of sweatshop workers took a back seat to the interests of business owners and consumers, who were delighted with the efficiency gains, reveling in a period of extraordinary growth, lower prices, and a greater diversity of clothing available to the United States' growing middle class.

The fire brought new urgency to the plight of industrial workers. The *New York Times* and the *World* pinned the blame squarely on the actions of a negligent, uncaring company, joining a chorus of activist newspapers that were already cheering progressive causes. They called for government action to prevent tragedies like this from happening again, putting pressure on politicians, such as New York governor John Dix, who had previously described himself as powerless to take action in response to the fire.

The unions turned the outrage into political power, organizing more than 90 percent of New York City garment workers by 1912. Politicians began to recognize that a "wait-and-see strategy" on workers' rights was not going to be sufficient and they called for investigations and indictments as the pressure grew. More than a decade after the ILGWU began to mobilize garment workers, the New York State Legislature established the Factory Investigating Commission with Frances Perkins as its chair. The commission was given "broad discretion in the investigation of fire hazards, as well as other factory conditions that adversely affected workers' welfare."

A flurry of progressive legislation on workers' rights soon followed. The commission secured the passage of thirty-six laws in New York State in a period of just three years, including on fire prevention, child labor, and working-hour limitations. In addition to a focus on working conditions, the concept of workers' compensation took off; eleven states passed workers' compensation bills in 1911 alone, and by 1948, every state provided the benefit. Decades after the emergence of sweatshops, it sadly took a tragedy for lawmakers to finally step up to address the dangerous conditions of sweatshop labor. When the social consequences of industrial activity become intolerable, government acts.

It's often said that the free market is the most powerful engine of innovation and human progress the world has ever seen. The idea contains much truth; decentralized marketplaces driven by entrepreneurs and private enterprise have been the best drivers of economic growth. Yet there is no such thing as a market completely free from rules and regulations. For one thing, public investments in universities as well as in industry are crucial to scientific discovery and technological innovation, including the basic building blocks of modern technology such as microchips and the internet. And the healthy operation of a dynamic marketplace depends on government for creating and enforcing rules of fair play, such as intellectual property and patents, antitrust, and consumer protections.

Nevertheless, markets work best when they are left ample space by government. This creates a difficult dilemma. On the one hand, policy makers are loath to get in the way of progress, and companies benefiting from an unfettered market are intent on preventing the government from slowing down the pace of change. On the other hand, democracies have objectives, such as protecting individual rights and ensuring basic safety and collective security, that go beyond economic growth. Our elected officials need to figure out how to achieve those objectives without sacrificing the benefits of a free market—or undermining the political support base that keeps them in power.

The result is a predictable dance among technologists making new discoveries, business owners, regulators, and ordinary citizens. The routine begins when human ingenuity and private capital spur extraordinary advances in technology that yield large-scale economic benefits. New companies, experimenting with the new technologies, proliferate. Over time, the effects of the new technologies ripple across society, the market consolidates, and people come to realize that the innovations have brought with them a set of problems, often negative consequences or concentrated market power, which puts other values at risk. Under pressure, governments then try to regulate the new technologies or industries in ways that address those harms but may undermine innovation or become quickly out of date, thereby diminishing their efficacy. This cycle underscores just how challenging is the task of regulators: technological advances are often scientifically complex; the social ramifications are hard to anticipate until, because of scandal or disaster, the harms are glaringly obvious; meaningful progress is difficult to achieve politically; and adjusting regulations is challenging once they've been adopted.

This same story has played out repeatedly in the history of government's relationship to telecommunications. Take the telegraph—the point-to-point text messaging system invented in the 1800s. The first commercial telegraph was constructed in 1839 in Great Britain, and by the 1850s the industry in the United States was intensely competitive, with multiple carriers serving identical routes. Though telegraphs were expensive to build, twenty companies had already laid over 23,000 miles of wire by 1852. But the profits from this new business were low, especially as the lack of integration across systems meant that carriers needed to keep investing in infrastructure in order to grow their customer base. The consequence was a period of consolidation in the United States during which Western Union emerged as the monopoly provider of long-distance telegraph service by the late 1860s. Though the federal government took tentative early steps to constrain Western Union's power, the legislation had limited practical effect and the government showed no appetite for more aggressive actions to rein in the company's domination.

According to Columbia University legal scholar Tim Wu, throughout the latter half of the nineteenth century, Western Union "was able to charge monopoly prices, support a newswire monopoly (the Associated Press) and discriminate against disfavored customers." This market power translated into political power as the firm used access to its newswire as a carrot and stick to shape the behavior of politicians. It wasn't until 1910 that Congress finally moved in a serious way, declaring that telegraph and telephone companies were what economists now call natural monopolies, with an obligation to offer their services at a reasonable price to all customers without discrimination. Nearly fifty years after the problems of Western Union's market power first began to surface, legislators finally got into the game. The "common carrier" provisions Congress adopted early in the twentieth century to prevent unfair discrimination—rules that prevented key players in the transport of goods, people, and information from exploiting their powerful position—remain the basis of telephone regulation today, even though so much about how we communicate with one another has changed in the meantime.

But even with a big legislative win, as technologies changed, the government struggled to stay on top of new developments. By the 1910s, as the telephone fully displaced the telegraph, AT&T achieved a commanding position in the long-distance market, acquiring local telephone companies to consolidate its position and denying interconnection to competitors. The threat of antitrust action by the federal government in 1913 led AT&T to agree to allow local companies to connect to its long-distance system. But with its economic and political power, the company developed workarounds, and enforcement was lax. AT&T's power and profits grew, and consumers were held hostage to a largely unregulated monopoly provider of telephone services. It took nearly two decades for the federal government to achieve a significant regulatory overhaul: the Communications Act of 1934. This legislation established the Federal Communications Commission (FCC), a new government body with responsibility for carrying out the federal regulatory role for both interstate telephony and radio. At its core, the new legislation ensured that local companies could access interstate connections and compete as local service providers.

Over the ensuing decades, the pace of technological change picked up while the regulatory structure remained the same. The introduction of microwave technology paved the way for new competitors in the long-distance market—an especially vulnerable market because the government permitted AT&T to charge high prices for long-distance calls in order to subsidize universal access. Cable services emerged, yet because cable didn't fit under the FCC's jurisdiction, it was subject to a confusing patchwork of local regulations and standards. With the high cost of the cable system's physical networks, concerns about market power arose in that area as well. By the 1980s and 1990s, the communications market was being flooded by yet more alternatives, including direct broadcast satellite (DBS) and futuristic models of "high-speed" data. The government's regulatory approach, designed for the age of Thomas Edison and long-distance telephony in the 1930s, was wildly out of date.

The race between innovation and democracy is real. Technologies are fast-moving. It is difficult to anticipate their effects, especially as they disrupt existing markets or create entirely new industries. Generating political consensus around what problems need to be solved and how to solve them isn't easy. And once consensus is achieved, adaptation is hard.

In the face of this dance, it's clear that our traditional choreography of regulation has its limits. In the words of the Nobel Prize–winning economist Paul M. Romer, "Finding good rules is not a one-time event." Our rules need to evolve quickly as new technologies arrive. They need to be responsive to increases in scale. And they need to be robust to the opportunistic actions of individuals and companies who will try to undermine them. Romer called this Myron's Law after another Nobel Prize–winning economist, Myron Scholes, who once commented in a seminar, "Asymptotically, any finite tax code collects zero revenue." By that he meant that clever people will find ways to work around anything that is fixed. The bottom line is that our rules need to be as dynamic as the technologies they are designed to regulate.

GOVERNMENT IS COMPLICIT IN THE ABSENCE OF REGULATION

The outcome we have today in the United States—a largely unregulated technology sector with growing market power that is harmful to individuals and society in ways that are impossible to ignore—isn't just a function of the ways in which innovation tends to outrun effective regulation; it also reflects the deliberate choices of democratically elected politicians in the 1990s.

The Clinton administration fashioned itself as quite forward-leaning when it came to the technologies of the future. Two young southern politicians, Bill Clinton and Al Gore, saw their time in office as a generational shift, one that necessitated remaking the Democratic Party and embracing the information communications revolution that was just beginning to get under way. Gore, in particular, was a champion of what he called the "information superhighway." Before becoming vice president in 1992, he was the chief architect of the 1991 High Performance Computing and Communications Act, which directed $600 million of funding for computing research and collaboration between academia and industry. Among other accomplishments, the legislation laid the foundation for the work of Marc Andreessen and others at the University of Illinois in creating the Mosaic web browser. In 1994, the Clinton administration created the first website for the White House, bringing to the world an online presence for the federal government far earlier in the World Wide Web's development than was the case for many companies and universities (at the end of 1994, there were fewer than ten thousand websites in the entire world). With an eye toward a digital future, Clinton and Gore anticipated an explosion in information services and believed that private capital and free markets were the key vehicles for driving innovation in this new space.

They pursued a wide range of policies consistent with that view. They kept the governance of the internet in private hands, rather than empowering government decision makers. They deregulated mobile telephone companies and conducted auctions for spectrum space to support the growth of wireless companies that would compete with traditional service providers. And they oversaw the passage of the Telecommunications Act of 1996, a watershed moment in the governance of the internet, which laid the foundation for the powerful and problematic technology sector we confront today.

At the core of the Telecommunications Act was a distinction between *telecommunications* services and *information* services. Such a distinction makes little sense today, as telephones, television, and the internet are virtually indistinguishable from one another. The powerful smartphones in our pockets contain all of these functions and many more on top of them. But at the time, the government was managing a legacy system of telephone companies while looking to accelerate futuristic communications technologies, such as broadband and the internet, which were still in their infancy. So the distinction between the two services could not have been more important. The old telephone system continued under the "common carrier" rules first adopted in 1910. Though steps were taken to promote competition in the telephone markets, the system still operated under government oversight, given the degree of market power and the need for universal access and regulated prices.

However, legislators adopted a wholly different framework for the new frontier called information services. At the time, the sector had major players such as America Online and CompuServe—companies that provided many of us with our first window into the internet—as well as private data networks linking computers and fax machines. Soon the sector would grow to include cable and broadband. Seeking to fast-track progress to the information superhighway, the Clinton administration excluded information services from traditional common carrier regulations. In effect, the Telecommunications Act was an open invitation to investors and companies to dive in. The chairman of the FCC at the time, Reed Hundt, spearheaded the pro-competition and deregulatory implementation of the bill, an approach to the emerging internet industry that would remain. As his successor at the FCC, William Kennard, said in 1999, "I want to create an *oasis from regulation* in the broadband world, so that any company, using any technology, will have incentives to deploy broadband in an unregulated or significantly deregulated environment" (author's emphasis). It's as if Kennard had happily heeded the words of John Perry Barlow. With assurances from the leading federal regulatory agency that its mission was dedicated to private competition and deregulation, a Wild West–style gold rush in Silicon Valley was on. As one measure of just how quickly the private market kicked into gear, it was only ten months after the Telecommunications Act was signed that Federal Reserve Chairman Alan Greenspan first warned about "irrational exuberance" in the stock market.

The decision to allow internet innovators to operate outside the public utility regulatory framework that governed telephone companies was the accelerant that sped up our arrival at the present moment. Telephone services, cable, and data are often bundled, and the distinctions among them are no longer relevant. But this approach also left unanswered critical questions that have come back to haunt us today, especially as we focus less on access and more on what happens to the content that moves across our communications networks. For example, must internet service providers and platforms treat all content equally as they decide what traffic moves and at what speed, just as telephone companies are forced to do under the old common carrier rules? Or can they make decisions based on what they like or don't like, their preferences reflecting editorial discretion, profit concerns, or some other motivation? The stakes here are enormous. These issues, bundled under the moniker of "net neutrality," are ultimately questions about how we are to manage market concentration and market power. The battle over net neutrality is ultimately a fight about whether your internet service provider has the right to speed up or slow down content passing through its network in order to make more money or favor particular providers.

Similarly, we now face complicated but hugely consequential questions about a notorious provision called Section 230 of the Communications Decency Act that was appended to the 1996 Telecommunications Act. With a few exceptions, Section 230 immunizes websites and internet service providers from legal liability stemming from any content posted by users. Whereas newspapers and television programs are content creators and therefore responsible for what they print or broadcast, internet service providers and social media companies can distribute user-generated content without incurring legal responsibility, even when that content is hateful, libelous, false, or vulgar.

Those making policy in the 1990s couldn't have anticipated that the flood of private investment they unleashed would generate both extraordinary innovation *and* extreme market concentration. A recent report from the New America Foundation on the cost of internet connection underscores the lack of market competition among internet service providers today. Not only has the United States fallen behind other developed countries in broadband penetration and internet speeds, but Americans also pay much higher prices. In 2020, Americans paid an average of $68.38 monthly for broadband, while the prices in France ($30.97), the UK ($39.48), and South Korea ($32.05) were significantly lower. The fact that other countries are doing better at delivering high-quality service and lower prices means that the problem isn't simply a function of the high costs of infrastructure in communications technologies but a reflection of policy choices in the United States.

For example, France's version of common carrier rules requires that the dominant service providers lease out the "last mile" of their network so that competitors have a chance to target consumers directly. But in the United States, a beautiful regulatory oasis, information services companies are free of that obligation. Instead of intense market competition and low prices, there has been ever greater concentration in telecommunications, with incumbents pushing their competitors out or simply acquiring young, promising start-ups.

It's no surprise that the dominant companies have turned their market power into political power, too, lobbying to preserve the core elements of the Telecommunications Act and seeking to prevent meaningful antitrust enforcement. In 2020, with antitrust actions against big tech companies finally in motion, the degree of their coordination to prevent regulation became apparent. In just one example, a government lawsuit revealed the extent of Google's partnership with Facebook, Apple, Microsoft, and Amazon to impede privacy legislation so as to protect its core ad business. So this isn't just a story of technology. Politicians are complicit in the market concentration we confront today, whether in telecommunications, where the service is spotty and prices are too high, or in any other domain—Amazon in e-commerce, Google in search, Facebook in social networks—in which the incumbent is reaping record profits and gobbling up its competitors.

Tom Wheeler, FCC chairman during the Obama administration, wryly observed that technology companies face a conundrum that Oscar Wilde first identified: "There are only two tragedies in life: one is not getting what one wants, and the other is getting it." Companies have "run the table" in Washington, positioning themselves to reap enormous private rewards from the growth of the internet while avoiding any of the public interest regulation that has typically governed the communications space. But with concern growing about tech's negative consequences and the extraordinary degree of market power in these industries, the era in which network and platform companies have made their own rules is coming to an end. The fiction that innovation simply outruns democracy is no longer tenable. Democracy shares the blame, and we need to figure out how to do better.

THE FATE OF PLATO'S PHILOSOPHER KINGS

During the summer of 2020, in the midst of the COVID-19 pandemic, an unprecedented event took place in Washington: in a virtual spectacle, lawmakers grilled the CEOs of the four major technology companies—

Amazon, Apple, Facebook, and Google—about allegations that they had abused their extraordinary market power. Representative David Cicilline, the chairman of the Subcommittee on Antitrust, Commercial, and Administrative Law of the House of Representatives, did not mince words in describing the issues at stake: "When [antitrust] laws were written, the monopolists were men named Rockefeller and Carnegie," he said. "Today, the men are named Zuckerberg, Cook, Pichai and Bezos. Once again, their control of the marketplace allows them to do whatever it takes to crush independent business and expand their own power. This must end."

It was the first time the four CEOs had appeared together jointly, and it was Bezos's first appearance before Congress. That in itself is shocking, given that Amazon already controls nearly 40 percent of the entire e-commerce market and concerns about its anti-competitive behavior have been voiced for years. The panel of inquisitors had done their homework, unlike in earlier hearings, when the elected officials had revealed a lack of understanding of the companies' basic operations. This time, they were armed with accusations of lawbreaking, backed by whistleblower testimony and internal company emails. The questioners demonstrated a well-informed understanding of the companies' underlying business models and evinced an eagerness to push for new forms of regulation.

One can imagine that the CEOs gathered their best and brightest in the C-suite to figure out how to counter the incoming criticisms. Amazon's Jeff Bezos and Google's CEO, Sundar Pichai, went for the personal approach—leading with their own inspiring stories of having risen from humble beginnings. Apple CEO Tim Cook continued his effort to differentiate Apple from the other companies, claiming that it doesn't have a "dominant market share in any market where we do business."

Facebook's Mark Zuckerberg took perhaps the most interesting line in defending the company's enormous market power: "Facebook is a successful company now, but we got there the American way: we started with nothing and provided better products that people find valuable." Zuckerberg underscored that Facebook had earned its dominant place in the market and should not be punished for its record of successful innovation. Suggesting that Facebook continues to face a competitive landscape, he argued that mergers and acquisitions are simply a part of Facebook's strategy to do better for its customers and that Facebook's size enables it to turn new products into extremely valuable services. He challenged the focus on market power, saying "As I understand our laws, companies aren't bad just because they are big."

As the CEOs and lawmakers fought it out in a stilted back-and-forth online, one couldn't miss the two competing views of the world. One is that technology is a source of enormous progress, a force for good in the world, and a source of economic, technological, and geopolitical strength and that more than anything else, the market power of these companies reflects their success in delivering high-quality products and services to consumers. From this perspective, if government gets in the way, it risks upsetting a virtuous cycle of competition and innovation. The other view is that network effects and an absence of regulatory oversight account for a nontrivial portion of the companies' market success. And what's good for the companies may no longer be good for all of us. As one commentator put it, perhaps we are coming to realize that "it is possible to love the services of Facebook and Google and question whether the benefits justify the harm." This points to the core issue we face today: Through our democratic institutions, charged with representing the interests of all citizens, can we preserve what's good about technology while eliminating or mitigating the harms?

Who is best situated to make these decisions? Should we trust the technologists, who argue that their market power isn't a problem but actually an asset that drives further innovation? Perhaps, as experts in technology, they are better positioned than typical Washington politicians with little technical expertise to know what kind of regulatory landscape is best for innovation. The politicians should just get out of the way. It's a totally reasonable argument to make and one that echoes a very long tradition of political thought.

Democracy has deep roots in ancient Greece. But the greatest philosophers of the ancient world were no proponents of democracy. Plato, whose dialogues form the foundation of Western philosophy, developed a blueprint for the ideal society in *The Republic* that entrusts its governance to a small group of skilled experts. In his lessons at the Academy, the first institution of higher learning in the Western world, he celebrated the rule of the wisest, arguing for the empowerment of enlightened philosopher kings—those with the "most intelligence of the principles that are the means of good government." He warned that liberty would give rise to tyranny as the city "intoxicated by drinking too deep of that unmixed wine [of liberty] . . . [would] finally pay no heed even to the laws written or unwritten." Democracy, or rule of the people, he felt, was a degenerate form of political organization. In the face of this horrific picture of life in a democracy, it is no wonder that his students were taken with the idea of entrusting the guardianship of society to a wise, capable, intelligent few. The most famous of those students, Aristotle, rejected much of Plato's philosophy, but he, too, thought democracy to be a deviant form of rule.

Over generations, philosophers have struggled with the tension between giving voice and decision-making power to citizens, which seems so appealing, and the recognition that experts of various forms are often better equipped to make good judgments on behalf of society. People have worried about whether citizens may be unduly influenced by momentary passions, embrace factionalism, and fall under the sway of demagogues— previously abstract ideas that have become all too real as a feature of today's democratic decay. The economist Bryan Caplan summed up the core problem in a memorable yet disturbing turn of phrase: "In my view, democracy fails *because* it does what voters want."

In a recent provocatively titled book, *Against Democracy*, the philosopher Jason Brennan tried to resuscitate Plato's vision of government by calling for the replacement of democracy with an epistocracy, or rule by experts. The author wasn't kind in his treatment of citizens, referring to them as "hobbits" (those who lack knowledge of politics) and "hooligans" (those who rabidly support one side regardless of the evidence). A better society, he argued, would be one in which we put "Vulcans"—such as *Star Trek*'s Mr. Spock—in charge. People who approach challenging political questions rationally based on evidence. People who can identify and solve problems. People who can deliver the best outcomes on behalf of society.

People who will, say, maximize the progress of science. People who are, in a word, optimizers.

One can see why this vision of rule by experts might be so appealing to technologists, given the unique and specialized knowledge they have access to. Since so few citizens or politicians even understand how our everyday technologies work, the case for an epistocracy—or technocracy—over emerging technologies seems strong. And it may be exactly what we need to navigate the "passions" of the masses who become fixated on one-off privacy breaches or viral doctored videos that risk distracting our collective attention from the wholesale benefits that technology has delivered. Representative democracies, with their know-nothing leaders, should stand aside.

But the idea of letting experts rule runs into a few serious problems. The first is determining who the experts would be. When Plato imagined a set of philosopher kings, he had in mind those who had special training, understood truth, and possessed skill in government with the means of balancing the interests of groups across the state. When Aristotle described the best forms of government, he was particularly focused on empowering rulers who would work for the common good rather than their own interests alone. What does it mean to have particular skills in government? At least in the modern variants of this argument, the idea of expert rule has been to privilege a particular mindset—the scientific mindset—which involves analyzing facts, developing recommendations, and making dispassionate choices consistent with evidence.

When technologists think about expert rule, they have something else in mind: either they would empower technologists themselves, with their privileged knowledge of what is required for innovation to thrive as policy makers, or their libertarian leanings lead them to favor Kennard's "regulatory oasis," that is, the best government is that which governs least, leaving technology companies free to make investments and design products unconstrained by other concerns. But the kind of expertise that technologists have is quite different from what Plato envisioned. Technologists have no unique skill in governing, weighing competing values, or assessing evidence. Their expertise is in building and designing technology.

What they bring to expert rule is actually a set of values masquerading as expertise—values that emerge from the marriage of the optimization mindset and the profit motive.

The second problem is one of legitimacy. For government to work, people must accept the decisions of those who rule. And legitimacy isn't earned only as a function of how effectively a government works. People want to involve themselves in the decision-making process. They want visibility into how decisions are made. They want to be able to challenge decisions if they find them unacceptable. They want government to be responsive to citizens, count their interests equally, and listen to their voices. On those counts, expert rule clearly falls short.

A third problem is the way in which expert rule entrenches existing holders of power and influence, because once a structure is in place that privileges the views of an educated, influential, and perhaps expert group, those who benefit from the arrangement have every incentive to maintain their status. If technologists write the rules—either as expert policy makers or as companies unconstrained in a regulatory oasis—we shouldn't be surprised if the resulting societal outcomes are ones that benefit technologists at the expense of everyone else.

There are no right answers to the questions we are struggling with. There are only better and worse answers. The answers we choose will reflect not only what the facts and evidence say but also what we value. In the words of the international affairs expert Tom Nichols, a passionate advocate for expertise in our public life, "Experts can only offer alternatives. They cannot, however, make choices about *values* . . . voters must engage those issues and decide what they value most, and therefore what they want done."

WHAT'S GOOD FOR COMPANIES MAY NOT BE GOOD FOR A HEALTHY SOCIETY

So the question is, do we want technologies, and therefore technologists, to govern us? Or do we want, through democratic institutions, to govern technology? When it comes to the regulation of technology, what does democracy have to offer? It's an important question to ask, especially at a time when democracy, this celebrated Athenian invention, seems to be in retreat.

The shelves in any bookstore chart our collective angst, with titles such as *How Democracies Die, How Democracy Ends, Surviving Autocracy,* and *On Tyranny* outselling their competitors. The social mobilization that followed the 2020 murder of George Floyd and many other Black Americans at the hands of the police surfaced longstanding concerns about whether communities of color can expect equal treatment under the law. The disastrous responses of some governments to the COVID-19 pandemic also appear inseparable from failures of democratic institutions, with the United States, Brazil, and India leading the world in deaths from COVID-19 and the European democracies not faring especially well, either. Even young people, the cohort we most associate with an embrace of liberty, are turning against democracy. One study found that 46 percent of Americans between the ages of eighteen and twenty-nine would prefer to be governed by experts than by elected officials. Another revealed that a quarter of millennials agreed that "choosing leaders through free elections is *unimportant*" (author's italics).

Arguments for democracy come in two forms. The first emphasizes the value of particular procedures for making decisions. If you believe that all people are entitled to basic rights and liberties and that they should be treated equally, there must be a way of making decisions across a diversity of viewpoints. It's obvious that democracies, despite their soaring rhetoric about liberty and equal citizenship, have long contained hierarchies or castes. Many of the world's major democracies embed discrimination along the lines of gender, race, and social class at their cores, despite the work of generations that have sought greater equality. Nevertheless, enthusiasts of procedural democracy, such as the nineteenth-century English philosopher John Stuart Mill, see democracy as the system best equipped to meet this challenge and overcome inequality. The reason is that democracies build in responsiveness to the interests of all persons. In Mill's words: "There is no difficulty in showing that the ideally best form of government is that in which the sovereignty . . . is vested in the entire aggregate of the community." This is especially true if we recognize, as Harvard professor Danielle Allen challenges us to do, that liberty and equality are inseparable from one another as part of building a democratic culture.

Others go even farther in celebrating the ways in which democracy prioritizes public deliberation and debate. Joshua Cohen, a prominent political philosopher who decamped from academia for the world of tech and now works at Apple University, pointed out that democracy does more than enable people to advance their individual views. Diversity of interests and values is a starting point of politics. If you give people freedom, they will choose to live in different ways, with different commitments, leading to reasonable disagreements about the best way of life. Mill called this "experiments in living," and he considered such diversity to be one of the chief benefits of liberty. But if we can reasonably disagree about the best way to live an individual life, how will we find common ground about how to live together in a single community? For Cohen, in debating questions of politics, "One must instead find reasons that are compelling to others, acknowledging those others as equals, aware that they have alternative reasonable commitments, and knowing something about the kinds of commitments they are likely to have." From this perspective, the very process of deliberation is valuable as it creates the conditions for persuading others of our views and establishing the possibility of a shared commitment to the institutions that collectively govern us.

The other view is that democracy beats the alternative of a nondemocratic regime not because of its distinctively fair processes but because it generates better outcomes, including innovation and economic growth. A democratic society that guarantees individual liberty and counts the interests of all people equally relies not on the expertise of its rulers but on the collective wisdom of its people. Democratic institutions enable us to harvest and aggregate the knowledge of our citizens. The free flow of ideas and debate is core to this argument, for without freedom, the advance of science is not possible. Mill argued that individuals need to be able to challenge the status quo and test out new ideas if innovation is to take hold.

An anecdote about Russian president Dmitry Medvedev's visit to Stanford University in 2010 underscores this point. While touring Silicon Valley, Medvedev wanted to understand what makes the innovation economy work. Proximity to a world-class university? Access to venture capital? The region's physical infrastructure? But although these might be things that a Russian autocrat could replicate, Medvedev was told that the region's success is tied to the people it is able to attract. And of course, one of Russia's problems is that many of the best and brightest Russian scientists prefer the freedom of Silicon Valley to the repressive environment of Moscow. Mill captured the value of liberty nicely when he wrote, "The general prosperity attains a greater height, and is more widely diffused, in proportion to the amount and variety of the personal energies enlisted in promoting it."

Democracy may also be more conducive to economic growth because it better protects the economic interests of capital holders and innovators. If government is good for anything, it should at least provide fair and stable rules of social cooperation and competition, as well as incentives for investment that will facilitate the growth of diversified and complex economies. Without economic growth, generating and spreading prosperity are difficult and it is hard to find the resources needed for education, health care, and social protection. One problem with autocrats is that they tend to view their country's economies as their personal piggy banks, extracting what they need to sustain their rule and enrich themselves and their families. Though it is undoubtedly true that some of the most compelling stories of economic growth in the last century occurred under autocratic rule—think China, Indonesia, and Chile—most nondemocratic governments have been development disasters, including those of North Korea, Zaire, and Zimbabwe.

Yet the abstract virtues of democracy are difficult to mesh with a few harsh realities that appear again and again when it comes to governing technology: the technical ignorance of elected politicians who are hardly credible in providing oversight or contemplating regulation; profound disagreements about what we value and how trade-offs should be made, whether in regard to data privacy, free speech, and content moderation or automation and the future of work; the slow, painstaking consideration of legislation that seems to generate competing bills—so that everyone has his or her name on one—without generating significant progress, especially in a highly polarized political environment; and the strong status quo bias of democratic institutions, which means that policy change is slow and sticky, making it difficult for regulators to respond flexibly and adaptably to new developments in technology.

The question of how much power to put into the hands of politicians when it comes to technology is one with real-world consequences for all of us. As just one example, when fourteen people were killed in a terrorist attack in San Bernardino, California, in 2015, a debate that had previously raged for years inside the halls of government burst into public view: Should the government have the authority to access the personal data contained on a cell phone, in this case the iPhone used by the alleged perpetrator? In the aftermath of an attack, it is obvious why access to this private information would be enormously useful to the police and federal authorities. But the tech companies had a different view: privacy, they argued, is a first-order value, and companies should be able to develop technology—for example, encryption—to ensure that no one, not even the government with a warrant, can access personal information. Jeremy saw such debates play out over years in government, where the White House Situation Room was a mirror of society's broader disagreements, with the technologists singing the praises of encryption and the national security policy makers puzzled as to why the technologists didn't seem to care about protecting Americans from terrorists. And here we are, still debating the same issue today. The coming chapters chart a number of domains in which issues exactly like this—where there are multiple, competing values—are flummoxing our system. And in the world in which we actually live, where many democratic institutions are polarized and paralyzed, the risks that more democratic oversight of technology will strangle innovation are not to be ignored, and the reality is that the technologists may simply outrun the rules.

DEMOCRACY AS A GUARDRAIL

Democracy offers the promise of freedom, equality, fairness, and deliberation, but its familiar and appealing ideals can also, in actual practice, be slow, uninformed, and ultimately constraining. If the decisions we confront in regulating technology do not have obvious answers—and most do not, for reasonable people can disagree how much free speech is good and what decisions should be kept out of the hands of robots—the idea that we will build some sort of shared and definitive view about how all good things and competing values can be reconciled is fanciful, especially in a world in which new technologies are coming online so quickly. So why should we continue to look to elected politicians to regulate technology, given their track record?

Part of the answer hinges on just how concerned you are about the consequences of the hands-off approach to new technologies. Silicon Valley's attachment to the motto "Move fast and break things" has generated real effects on how much privacy we enjoy, the nature of work, and what we are exposed to in the digital public sphere. We have called these effects externalities—the by-products of technological change and innovation—and it falls to government, if it can get its act together, to deal with the consequences.

Our view is that these consequences are far more significant than the developers of these technologies sometimes anticipate. We also shouldn't wait until we experience the unintended consequences before beginning to think hard about how they might be mitigated or addressed. We can do better than sticking with a mistake-driven culture in which we look to government to react only when things go horribly wrong.

If Winston Churchill was right that democracy is nothing more than "the worst form of Government except for all those other forms," it is worth focusing on the minimal yet fundamental task at which democracies seem to excel: avoiding catastrophic outcomes and striving for stability and resilience to unpredictable shocks. Because even if we can't agree on exactly the kind of society we want to live in, we can typically come to a consensus on the worst outcomes we want to avoid, such as imposing harm on individuals, being cruel to disadvantaged populations, and creating second-class citizens.

Amartya Sen, a Nobel Prize–winning economist at Harvard, provides a contemporary demonstration of democracy's value as a guardrail protecting societies from bad outcomes. He drew attention to the striking observation that no democratic country has ever experienced a famine. This is a puzzling phenomenon if one thinks that famines are natural disasters, caused by changes in weather and climate. Sen showed that they are in fact man-made political disasters, failures of government to organize and distribute sufficient food to the parts of a country that are subject to extreme weather shocks. Here the virtues of democracy come into clear view, because for Sen, the real value of democracy is the fact that elected leaders must in some sense be responsive and accountable to citizens. If you are starving to death, you will make your voice heard, or others will do so on your behalf. And in a world in which nothing is more fundamental than access to food, politicians know that a failure to avoid famine is sufficient cause for defeat at the ballot box. So they work harder—and harder than those in nondemocratic regimes—to avoid this worst-case scenario. Democracy, it turns out, is a kind of man-made technology that eliminates famine.

The often-incompetent responses of democratic governments, including in the United States, to the COVID-19 pandemic might cause you to question Sen's conclusion. But as we saw in the debates about the COVID-19 response during the run-up to the US election, the policy landscape was a battle over competing values—whether to prioritize the economy or public health, to protect only the most vulnerable or to reduce exposure for all Americans—and politicians were vying for voters based on these alternative visions.

The idea that democracy can help us avoid really bad outcomes has a long history in political thought. Perhaps its most vocal proponent was the twentieth-century Austrian philosopher Karl Popper, who was frustrated by the "lasting confusion" Plato had created in political philosophy. By focusing on the question of "Who should rule?," Plato stacked the deck in favor of an answer—the best, the wisest, he who masters the art of ruling—that fit his preferred worldview. But who would argue for anything else? Popper asked. Who would argue for the rule of the worst?

The right approach, in Popper's view, is to prepare from the beginning for the possibility of bad government and to ask: "How can we so organize our political institutions that bad or incompetent rulers can be prevented from doing too much damage?" This shifts our attention away from trying to find the best, most expert leaders to creating rules and institutions that enable us to remove bad leaders and reward good leaders when we have them. The bottom line is that we need good rules, not simply good rulers. And good rules are not a fixed point once they are deduced; they need to adapt to changing social and economic conditions, including technological innovations. And so democracies, which welcome a clash of competing interests and permit the revisiting and revising of questions of policy, will respond by updating rules when it is obvious that current conditions produce harm; when people suffer needlessly or unfairly; when a fire in a sweatshop awakens us to a long-simmering problem.

It thus falls to all of us—the citizens—to work the system to get the outcomes we want. In plain terms, Popper said, "It is quite wrong to blame democracy for the political shortcomings of a democratic state. We should rather blame ourselves, that is to say, the citizens of the democratic state."

This matters for the governance of technology because it is a deep critique of the utopian notion of social engineering—the idea that we can organize our politics to arrive at the one best outcome for society. Such a thing is not only unrealistic, it is also dangerous because it creates a pathway straight to dictatorship.

We need an alternative model of what we want from our politics—not utopian social engineering, as Plato imagined, but "piecemeal engineering," in Popper's words. Or in the words of another twentieth-century philosopher, Judith Shklar, we want a democratic society that rejects "a *summum bonum* toward which all political agents should strive," but that does begin with "a *summum malum,* which all of us know and would avoid if only we could." Without a blueprint for where we want to end up—because such a thing is impossible to achieve—we must focus instead on identifying and mitigating the harms and suffering we want to avoid. At this task, democracies have generally excelled: avoiding mass starvation; preventing nuclear war; eliminating extreme poverty and suffering.

This is a pretty minimalist view of what democracy is good for and is nothing to apologize for. Like a famine, the effects of technology on society are a man-made disaster: we create the technologies, we set the rules, and what happens is ultimately the result of our collective choices.

Tom Wheeler, the former FCC chair, likened the present moment to the progressive era with which we opened this chapter. "In a time of rapid technological change," he wrote, "innovative capitalists step up to make the rules regarding how their activities impact the rest of us." But then, he continued, "such self-interested rule-making has been confronted eventually by a collective public interest, democratically expressed, to create new rules that protect the common good."

Our challenge is to determine just what that common good entails and how we can harness our democracy to achieve it. This requires focusing on the technologies of the future and the opportunities before us to chart a different way forward.

PART II

DISAGGREGATING THE TECHNOLOGIES

What the inventive genius of mankind has bestowed upon us in the last hundred years could have made human life care free and happy if the development of the organizing power of man had been able to keep step with his technical advances. . . . As it is, the hardly bought achievements of the machine age in the hands of our generation are as dangerous as a razor in the hands of a three-year-old child.

—Albert Einstein, writing in the *Nation*, 1932

CHAPTER 4

CAN ALGORITHMIC DECISION-MAKING EVER BE FAIR?

In 1998, on the heels of a few new acquisitions and a modest IPO, Amazon CEO Jeff Bezos set out to articulate the company's core values. He identified five, including a "high bar for talent." Though the company was still in its early days, he knew that attracting a high-performing team would be essential to realizing his grand vision of an everything store, something that far exceeded the company's then identity as "the world's largest online bookstore." In the two and a half decades since, Amazon has surpassed all expectations. It has entered countless new markets, transformed the customer experience of online retail, and become the second publicly traded US company to be valued at more than $1 trillion. As its valuation has soared, so, too, has its workforce, ballooning from 614 employees in 1998 to a global full- and part-time workforce of more than 750,000 today. On any given day, the company hires an average of 337 people and has nearly 30,000 open positions.

In light of this expansion, an obvious question has emerged: Is it still possible for Amazon to maintain the high bar for talent that Bezos projected in the company's early years? Amazon HR chief Beth Galetti believes it is—and that the same innovation that fueled the company's meteoric rise might also drive its ambitious approach to talent. "If we're going to hire tens of thousands—or now hundreds of thousands—of people a year," she has said, "we can't afford to live by manual processes and manual transactions."

It was in this spirit that in 2014, Amazon began putting some of its technical horsepower to work on a new challenge: recruiting and hiring top talent. The company's engineers envisioned a new tool that could use algorithms to identify the most promising candidates. With powerful machine learning techniques that they hoped would identify the best of the best, they planned to train the new system using the previous ten years of résumés the company had received, as well as other internal data that could make the model more precise. Over time, the system would learn to recognize the qualities, skills, credentials, and experiences that make applicants successful at Amazon. Candidates would be given a score between one and five stars based on potential, just as consumers rate products on the retailer's platform.

The promise of this tool was clear and its purpose compelling. If Amazon could dramatically enhance its hiring process through smarter automated tools, it could increase the efficiency of its recruiting operations and double down on its long-standing commitment to a "High Bar for Talent," all while maintaining extraordinary growth in the core business. Moreover, a human review of tens of thousands of applicants' résumés per year is costly, and just as Amazon strives to economize on price for its customers, the deployment of algorithms in HR could provide significant cost savings. According to one source from the company, "Everyone wanted this holy grail. They literally wanted it to be an engine where I'm going to give you 100 resumes, it will spit out the top five, and we'll hire those."

In addition to the efficiency gains, moving from a process driven by human judgment to one driven by algorithms and data offered Amazon an exciting possibility: to build a recruiting system that would be free of human bias—or at least would improve upon the bias-riddled decision-making of humans. For years, researchers have shown that racial and gender discrimination routinely play a role in hiring decisions and that human decision makers are plagued by a bevy of other biases both conscious and unconscious. When identical résumés were circulated to potential hiring firms using different names (e.g., a very African American–sounding name or a very white-sounding name), results consistently showed significant discrimination based on perceived race, with white names receiving 50 percent more callbacks for interviews. In building a new hiring tool from scratch, Amazon could strike a blow for social justice by liberating itself from the historical biases humans accumulate through their lived experience. It could make more precise, more efficient, and more objective hiring decisions—an achievement that would be meaningful at any time, especially as Galletti was gearing up to triple the company's workforce yet again.

But as recruiters began looking at the recommendations generated by the new system, something seemed amiss. The scores appeared to have a strange bias against females and a strong preference for males. As the team examined the findings more closely, they discovered that the algorithm had not just learned neutral patterns to predict an applicant's future job success but was amplifying a preference for male candidates it learned from the company's historical hiring data. As it turned out, the algorithm penalized résumés that included the word "women" at all, taking note of everything from "women's soccer captain" to "women in business," and it downgraded applicants from women-only universities. The engineers were not sexist. They had not deliberately inserted that bias or actively programmed a "sexist algorithm." Yet gender bias had crept in. The team attempted to adjust the code to neutralize the bias, but they could not rid the tool of all potential discrimination. After years of effort, Amazon decided to scrap its vision for the tool altogether and disband the team responsible for it.

The Amazon case is a revealing one and sparks specific questions we should ask about the rise of automated decision-making tools: If one of the most powerful companies in the world can't successfully build an algorithmic tool free of bias, can anyone? When new technologies are deployed to aid in or replace human decision-making, what standard of objectivity should we hold for their automated replacements? Who should be held responsible for troubling decisions made or informed by an algorithm? And who should decide whether or not to use these new tools in the first place?

WELCOME TO THE AGE OF MACHINES THAT LEARN

Algorithmic decision-making models are built using machine learning, which is essentially the process of finding patterns in data. Want to determine who should be interviewed for a job? First, get lots of data in the form of résumés of job applicants who already have been interviewed and note which ones were eventually hired and which ones were not. Then feed the data into a machine learning algorithm, which uses optimization to find the patterns—for example, determining important phrases on résumés—that best distinguish those who were hired from those who were not. Through determining these patterns—a process called "training"—the algorithm produces a model that can then be used for decision-making. The model has learned these distinguishing patterns in the data by trying to optimize some criterion, such as predictive accuracy—that is, how often it would make the correct hiring decision on the résumés of previous job applicants it was given. When training the machine learning model, the algorithm makes adjustments that lead it to make fewer and fewer mistakes. Critically, the choice of what criterion to optimize is made by the programmer, who could choose, for example, to simply have the algorithm screen out the résumés of applicants who are clearly unqualified or to push the algorithm to try to make the more nuanced decision of who should be hired.

Such model adjustments might include changing how particular words or phrases found on the résumé should be weighted as indicators of employability. For example, if we're trying to train a model for screening product manager applicants, the model might learn that the term "MBA" or more specifically "Wharton" or "Harvard Business School" should receive high weighting, while the term "truck driver" should receive a low or negative weighting. In more complicated models, the algorithm might look for combinations of words and phrases that are even more meaningful when taken together, such as finding the terms "founder," "funding," and "millions" together on the same résumé.

When the model has achieved a suitably high accuracy rate in making predictions, it is ready to use. Present the model with a new résumé, and it will make a prediction and potentially provide a confidence score for whether or not a given person should be hired. Take the list of people that the algorithm says should be hired, sort them by the algorithm's confidence scores and—voilà—you have an algorithmic way to decide who should be interviewed or offered a job, bypassing interviews with humans altogether, as some companies are now doing.

Of course, programmers can add constraints to the model. Say they don't want the model to infer anything from the terms "men" or "women"—as in "men's soccer team" or "women's engineering club"—because they want to avoid gender bias. That might seem reasonable enough. But some terms such as "baseball" or "softball" that the model can still consider are highly correlated with the applicant's gender and can lead the model to make decisions that exhibit gender bias. That's the kind of situation Amazon found itself in with its résumé-screening tool.

Moreover, as job applicants become aware that automated tools are used for screening their résumés, there's no end to the ways they can game the system. Say you're an applicant and you know your résumé will be analyzed by a machine. You could submit an electronic copy with some extra text in a white font on a white background at the bottom of the page. Since the text is white, it's invisible to any human who reads it online or prints a copy. But the automated résumé-screening tool will scan and process all those terms just as if they appeared in black ink on a white page. That "extra" text might just include every phrase that you think might give you an edge for the position you're applying for, including the entire list of required and desired applicant attributes in the original job posting. To stand out even more, why not add the names of an assortment of prestigious universities: Harvard, Oxford, Berkeley. Or signal something about your social position by including extracurricular activities such as "equestrian club" and "squash team." If you think that sounds far-fetched, think again. These are all examples shared with us by current and former students. This kind of gaming has been going on for years among the tech savvy, and it's only likely to get worse as algorithmic decision-making tools become more broadly used.

Proclamations about how machine learning is making computers "smarter" than humans are commonplace in the media these days. It hasn't always been that way. In fact, machine learning as an academic field has existed since the 1950s, when a researcher by the name of Arthur Samuel first programmed a computer to get better at playing checkers as it played more games and observed which actions led to wins versus losses. The computer quickly became good enough at the game to beat its programmer. But it's only in the past few years that this technology, often referred to by the more general moniker *artificial intelligence*, has come to the public's awareness. So why has a field that's more than half a century old suddenly become the cause célèbre among tech companies and the media?

Three things happened in the last decade that moved machine learning from the academic research lab to the point where Russian president Vladimir Putin declared, "Whoever becomes the leader in [artificial intelligence] will become the ruler of the world." First, computers became faster. Much, much faster. And they were networked in "the cloud," so that rather than having to do computation with a single machine, thousands of computers could be orchestrated in a computational symphony to solve really big problems. Second, the amount of digital data that was available grew by leaps and bounds. As more and more people shopped online, clicked on ads, liked their friends' social media posts, uploaded family photos, accessed their medical test results online, and went about their merry way on the internet, they were leaving behind a data stream constituting a treasure trove that could be used to learn about their interests, behaviors, preferences, and much more. Third, the researchers working on machine learning developed more powerful algorithms that could make use of the vast increases in computing power and data to build much more accurate—and much more complicated—models to make predictions about everything from which movie someone might like to whether that person might have a mental health issue. Train a learning algorithm by giving it images that contain faces versus those that don't, and you can create a model that can identify faces in any new image. Train a learning algorithm using X-ray images on which cancerous versus noncancerous regions have been identified by expert doctors, and you can build a model that can potentially predict whether cancer is present in a new X-ray. The possibilities are endless.

Compounding these advances was the development of clever new techniques that made it possible to harness even greater volumes of data to feed to machine learning algorithms. Traditional machine learning methods for distinguishing strong candidates from weak ones, for example, required that each résumé used to train the model be tagged by a human as to whether it was from a person who had been hired or not for a particular job. Similarly, facial recognition required human supervision to indicate if and where a face existed in each picture. Data that are labeled in this way are called "supervised" data, as they require human supervision to tag the data with a label that the model is trained to predict. Of course, there are millions more résumés available online that no one has tagged and billions of family photos sitting in online photo albums and on social networks that haven't been tagged.

Researchers eventually found effective ways to unlock the potential of such unlabeled—or "unsupervised"—data. The trick is to build a model with the small amount of supervised data available and then use the model to predict labels for a large volume of unsupervised data. Now armed with a new batch of labeled data, programmers repeat the process over and over, thereby enabling them to label even more of the previously unlabeled data. Rinse and repeat with oceans of unlabeled data points that are just sitting out there on the open web or have been collected by companies such as Google and Facebook that track nearly everything we do online. Many thousands of computers then process all of this at high speed. A revolution in capability was unleashed that is growing to this day.

Of course, as the pool of labeled data expands in this way, it is also possible for the model to go horribly wrong since an erroneous prediction early in the process will be compounded by later predictions. An African American software developer, Jacky Alciné, took to Twitter in 2015, for example, to take Google to task when its Photos application labeled pictures of him and his girlfriend as "gorillas." An engineer at Google quickly issued a public apology and vowed to fix the issue, replying on Twitter, "Holy fuck . . . This is 100% Not OK." But the damage was already done, and repairing the problem was no easy task. For several years afterward, Google's solution was to eliminate all gorillas and chimpanzees from its image bank. Machine learning has great potential both to solve complicated problems and to make disastrous mistakes.

We've come a long way since the 1950s, when a machine first learned to play checkers. A computer with the ability to beat an amateur player in checkers doesn't raise an eyebrow today. But a computer with the ability to diagnose your cancer better than a trained physician—a development that's actually happened in the past few years—makes headlines. In many ways, the rise of machine learning is an inevitable result of the ever-increasing speed of computers and the shift of so much human activity into the digital domain. It only makes sense that this technology, combined with an optimization mindset, would develop new ways in which humanity can improve on the error-prone decision-making processes of biased and inconsistent people. Of course, one major concern is that the data that algorithms are trained on often come from those same error-prone, biased, and inconsistent people. If the résumés given to a machine learning algorithm are flagged as "hire" or "do not hire" by humans, the algorithm happily learns the patterns in *human* decision-making—faulty or not—that generated those data.

It would be natural to wonder how machine learning models could be produced that are *better* than human decision makers if all they're trying to do is mimic human decisions. The answer is that in many domains, data are available that do not involve human judgment. Consider the decision of whether to grant bail to a defendant awaiting trial in the criminal justice system. The goal is not just to have the algorithm mimic the decisions of human judges, which studies have shown can be highly variable and error prone. Rather, the data that are used to train the algorithm include defendants who were released on bail and whether they subsequently appeared for their trial or not (or perhaps whether or not they committed another crime while out on bail). The algorithm thus learns which of the defendants' characteristics—based on their prior criminal record, the current charge they are facing, and a host of other factors, such as whether they are employed, married, have children, and others—may make them likely to appear for their trials or not. No human judgment is required or, perhaps more important, desired.

Moreover, algorithms created in this way deliver uniform results. If a courthouse in New York and another in Alaska use the same algorithmic tool, each will get the same risk score. This is a great way to eliminate human bias from high-stakes decision-making not just in criminal justice but in many different areas, such as who should be given a mortgage or what medical procedures should be approved for a patient.

DESIGNING FAIR ALGORITHMS

If the failed Amazon experiment had been an isolated example of algorithms gone awry, there would be no reason to worry. But the problems with the hiring tool at Amazon are not unique. We need to pay attention to such failures because algorithms are at work in so many different parts of our lives in ways that we don't even realize. In addition to corporate hiring, algorithms help determine our romantic lives when we use an online dating service; the health care we receive (or don't) and the price we pay for it; whether we qualify for a loan; whether we're eligible for housing or welfare benefits; what we see online; and what we learn in school. They provide early warnings of mental health problems; identify potential tax fraud; help decide whether a defendant will go to jail or get out on bail; determine the length of criminal sentences; and determine whether someone is eligible for parole. These are among the most important areas of any life: love, work, health, education, finance, and opportunity. Algorithms are also behind the targeted ads we see online, the responses to which drive the business model of many tech companies.

Even when algorithms are working well, we have to consider a wide array of important questions that go beyond the technical issue of their predictive accuracy.

Imagine yourself in the shoes of Eric Loomis, a thirty-four-year-old from Wisconsin who in February 2013 was caught driving a stolen car that had been used in a shooting. He pled guilty to evading arrest and no contest to operating a vehicle without the owner's consent. Neither crime required prison time. At his sentencing, however, the judge relied on an algorithmically generated risk assessment called COMPAS that indicated that Loomis was highly likely to reoffend. The judge rejected Loomis's request for probation and gave him a six-year prison term instead. Neither the judge nor the lawyers, and certainly not Loomis, understood how COMPAS works. They received only the output from the algorithm: a risk score. Northpointe, the company that produced the technology and had sold it to the state of Wisconsin, refused to divulge the algorithmic model, treating it as intellectual property. When Loomis's lawyers sought to appeal his sentence, they demanded an explanation for the risk score he'd received, an explanation that no one could provide. So Loomis sued Wisconsin for violating his due process rights, which, he believed, should afford him the right to challenge any evidence used against him in court. Wisconsin's Supreme Court rejected his challenge, placing him and other criminal defendants in Wisconsin in a Kafkaesque scenario of being denied an explanation for something as high stakes as whether they are sent to jail. Just as Amazon's algorithmic hiring tool violated our sense of fairness, so, too, does the COMPAS risk assessment.

It's more important than ever that engineers design algorithms that are fair and, better yet, make our world fairer. Some have taken note of the problem and generated a new line of academic work devoted to the topic. One hyperconfident website, called Spliddit, offers up what it calls "provably fair solutions" to problems such as sharing rent payments, assigning credit for group tasks, dividing estates between heirs, and splitting up chores or work shifts among a group of people. The nonprofit venture trumpets "methods that provide indisputable fairness guarantees." Buried deeper in the site, the creators explain, "When we say that we guarantee a fairness property, we are stating a mathematical fact." These computer scientists appear to believe that fairness is reducible to a mathematical formula. If only achieving fairness were so simple!

A separate group of scholars has reached a different conclusion. They've witnessed problems with algorithmic decision-making and created a group called FAccT/ML, an abbreviation for Fairness, Accountability, and Transparency in Machine Learning. Their stated goal is to help ensure that algorithmic decisions do not create discriminatory or unjust impacts across different demographics such as race, sex, religion, and others.

It would seem to be a promising development. But a deep problem presents itself immediately: how to define fairness. To highlight that point, one presentation at the group's 2018 annual conference provided twenty-one distinct definitions of fairness, each of which implied a different mathematical formulation of the concept. One definition states that algorithms should be gender blind—no gender identifications should be used as inputs—in order to be considered fair with respect to gender. Another requires that algorithms use gender as an input in order to overcome historical bias against women. Still another definition notes that fairness implies that the percentage of errors made by a programming model for women should be the same as that for men. So if a model is used for screening résumés, the percentage of women who are incorrectly identified as "not a hire" should be the same as the percentage of incorrect predictions for men. Other researchers have shown that several common specifications of fairness are incompatible in that trying to maximize one version might lead to lowering another. The upshot is that fairness is not easily understood as a timeless and universal thing on which we all agree. Instead, we must pay attention to what fairness means in particular social contexts.

Most people think there is a common understanding of *fair* and *unfair*. Yet fairness resists easy definition. Consider the following example. A school district is trying to decide how to finance its schools in order to best educate all the children in it. Someone says that fairness requires treating every child the same. By this definition, all children should get the identical amount of funding. To give unequal funding—to spend more per pupil, for example, on boys, or white children, or those who are religious, or those who are native born—would be to discriminate against girls, minorities, agnostics, and immigrants. And that would be unfair.

Then someone points out that some of the children who attend school have special learning needs. Some kids are dyslexic, blind, or deaf. Others have cognitive or physical impairments. To provide these children with the special education services they will need in order to learn, especially if they are to have an opportunity to learn that is comparable to that of children who are not blind or deaf, will cost more money. Specially trained teachers will have to be hired, special equipment will need to be purchased, classroom accommodations will have to be made. So isn't it only fair to spend more money on a per pupil basis for children with special needs?

So which is it? Does fairness require the same treatment or different treatment? Both views of fairness have reasonable merit.

Such disagreements about fairness are a routine part of our personal lives, too. As a parent, you try to treat your children fairly. But what does that mean in practice? If you want to give each child the chance to learn a musical instrument and one child wants to play the guitar and another the piano, will it strike you as unfair when you realize that a piano is much more expensive than a guitar? Do you give your children equal or different allowances? Perhaps it matters how old they are, but when you give cash gifts to your adult children, do you give them equal or different amounts? It's not obvious what fairness demands in these cases.

Complicating things still further, fairness is an idea that we apply to individuals as well as to groups. When we think about a hiring algorithm, we can construe fairness as a property of individuals, so that all applicants with identical skills and experiences get the same predictive score when they are reviewed by the algorithm. We can also construe fairness as a property of groups so that the percentage of minority-group members who are rated as hirable is the same as that of the majority group. Both conceptions of fairness seem important and reasonable. Yet we can't easily adopt both at the same time when designing algorithms.

So fairness in the design of algorithms is not easy to define or implement. It is contextual, and it depends on our collective social understanding about the circumstances involved.

Despite this, all is not lost. Fairness may not be consistently reducible to math, and it may vary somewhat across different social contexts, but this does not mean that it is subjective. We can still find our way and make use of this philosophical ideal. Indeed, it's inevitable that we do so, as fairness appears to be evolutionarily wired into human interaction. Sit down at a table of kindergartners, and offer them some stickers or candies. If you give more to some than to others, the children, especially those who receive less, will complain. The ones who receive more may not give theirs up, but they will recognize the unfairness. Studies in many different countries show that children even as young as twelve months have an understanding of fairness, exhibiting anger when they are treated differently by their peers, parents, or researchers. They display a strong aversion to inequitable outcomes and will threaten or sanction those who refuse to share in a fair manner.

Then there's the astonishingly consistent result from what researchers call the ultimatum game. Performed in dozens of countries with widely varied participants, the ultimatum game is so simple that you can try it with your friends. One person, the proposer, is given a sum of money, say one hundred dollars, and is told to share it in any manner she wishes—fifty/fifty, fifty-five/forty-five, one hundred/zero—with a second person, the responder. The responder has veto power and can accept or reject the deal as proposed. If he accepts, both walk away with the proposed split. If he rejects, both receive nothing. The proposer needs to figure out what share of the money she needs to offer in order for the responder to agree. A common idea in economics is that it is rational for the responder to accept any offer, since walking away with something, even one dollar, is always better than walking away with nothing. Yet responders everywhere tend to reject unequal offers and nearly always reject deeply unequal offers, sacrificing a material benefit for themselves in order to punish the proposer who exhibits unfairness. The widely accepted conclusion of these studies is that humans have a deeply rooted instinct toward fair treatment.

Even other species appear to have a strong fairness norm. A famous study by Sarah Brosnan and Frans de Waal used capuchin monkeys to test the idea. Two monkeys sat in adjacent cages, and a caretaker offered them a piece of food in exchange for doing a simple task. When the caretaker offered both monkeys cucumbers in exchange for the task, the monkeys complied happily and ate the cucumbers. But when the caretaker offered one monkey a grape—a much sweeter and preferable food—and the other monkey a cucumber, the disadvantaged monkey rebelled, rattling its cage and hurling the cucumber at the caretaker. Primates, Brosnan and de Waal concluded, are hardwired to resist certain kinds of unequal treatment.

When thinking about algorithmic decision-making, it is helpful to distinguish between two kinds of fairness: substantive and procedural. Substantive fairness focuses on the outcomes of some decision. Procedural fairness focuses on the process that generates the outcome. If the process is deemed fair, we don't have to concern ourselves with the outcome. Algorithms that are fair will involve both substantive and procedural considerations.

For fairness, the most important issue is determining what counts as a morally relevant consideration in the decision-making process. The oldest definition of fairness, going all the way back to Aristotle, is that fairness means treating like cases alike and different cases differently. The color of your hair is obviously morally irrelevant when it comes to whether or not a hiring algorithm should recommend that you be hired, a criminal justice risk score should recommend that you be jailed rather than released on bail, or a school district should provide identical per pupil funding. These are easy cases. What about gender, race, and religion? Are these characteristics morally relevant or irrelevant when we think about matters of great concern such as hiring, criminal justice, and educational opportunity? To answer this question, we can consult theories of justice for an answer, or the law, or our moral conscience. What's more, our understandings evolve over time and can differ from one social context to another. The US Constitution began by counting slaves as three-fifths of a white person and disenfranchising women. Differences between blacks and whites and between women and men were thought to be morally salient; long social struggles were necessary to overturn such differences in the law and are still ongoing with respect to differences in the minds of ordinary citizens. Today in the United States race is what lawyers call a "suspect classification," a characteristic of persons that can generate discriminatory treatment. Whether such a characteristic is morally relevant or irrelevant is what makes fairness a difficult ideal.

This difficulty is what gives procedural fairness its appeal. Since what counts as a substantively fair outcome is debatable, perhaps a focus on fair processes can rescue us. Do you need to divide up a birthday cake in a fair manner and give each person an equal slice? The fair approach is to give the knife to one person to divide the cake up and have him or her choose last. But this ostensible fairness doesn't totally eliminate the need to make decisions based on substantive fairness. A person on a diet might prefer a small piece of cake; someone who hasn't eaten all day might like a larger one. If these are relevant considerations, a fair process alone won't deliver.

Focusing on procedural fairness, however, can illuminate especially important considerations when it comes to thinking about the justice of any decision-making process, including algorithmic decision-making. The late philosopher John Rawls developed a theory of justice that he called "justice as fairness," where a fair decision-making process would not permit the powerful to take advantage of the weak, the rich to dominate the poor, or a majority group to outvote a minority group. To capture this mindset, he suggested that decisions impacting society should be made from behind a "veil of ignorance," where decision makers would have no knowledge of their own personal situation or socioeconomic status. Thus, they would have no motivation to make decisions to benefit their own personal interests (of which they would have no knowledge). Rather, their only goal would be to make decisions that would benefit society overall, paying particular attention to the people who might be most adversely affected by the decision, since they might be one of them without knowing it.

When it comes to algorithms, Rawls's veil of ignorance and emphasis on a fair procedure can be useful. If you were subject to some algorithmic decision-making procedure, say a decision about whether you deserve medical treatment or whether you will be recommended for hire from a pool of thousands of applicants, what would you want to know about the algorithmic model to give you confidence that it would treat you fairly? You'd want to know that the model had not been trained on data riddled with human bias with respect to things that are irrelevant. You'd want to understand how the model works—what factors it takes into account and which it ignores. You'd want some assurance that the programmers of the algorithm had not consulted merely their own intuition about what counts as morally relevant or irrelevant but were coding with a broader social understanding in mind. Faced with the black-box decisions of an advanced machine learning algorithm, whose output can't easily be explained even by the programmers, you'd have questions about fairness. And you would deserve answers.

ALGORITHMS ON TRIAL

The advance of algorithmic decision-making is rapidly outpacing our ability to agree on what we want these systems to achieve. Amazon's hiring algorithm is just the tip of the iceberg in the business world. And the enthusiasm for algorithms isn't limited to the private sector. Governments are getting into the game, including in areas you might have thought should be off limits to machines, such as how to allocate critical social services, whether to remove a child from a home, and what to teach children in school. In each case, someone has to make a decision about what the algorithm should optimize and how fairness is to be achieved—and these decisions are virtually invisible.

It is in the domain of criminal justice that the debates about fairness are most hotly contested. Take California as an example. In August 2018, to great acclaim, Governor Jerry Brown signed into law a sweeping overhaul of the state's bail system. The law eliminated cash bail in an effort to ensure, in Governor Brown's words, that "rich and poor alike are treated fairly." Under the new law, local courts would be asked to decide whether those arrested and charged with a crime should be kept in custody or released while they await trial. For nonviolent misdemeanor cases, the default would be release within twelve hours. But in other cases, the decision would be based on an algorithm created by courts in each jurisdiction to rate individual defendants on how likely they are to show up for their court date, the seriousness of their crime, and the likelihood of recidivism.

Those behind the legislation were motivated by the unfairness of the old cash bail system. It's easy to see their point; it amounted to systematic discrimination against the poor and disadvantaged. The decision about whether to detain someone before trial should presumably be made on the basis of the person's risk to the community rather than his or her ability to pay bail. Assemblywoman Lorena Gonzalez put it this way: "Thousands of sex offenders, rapists, and murderers are let out because they simply have money, [yet] somehow we are safer by keeping that system?" The bill's cosponsor, Senator Robert Hertzberg, called it "a transformational shift away from valuing private wealth and toward protecting public safety."

The bill's enthusiasts were also driven by evidence that an algorithmic approach to pretrial detention is actually better than one that relies on judges to make such decisions, in that it reduces the number of crimes committed by those awaiting trial while decreasing the number of people who are incarcerated before they are convicted. But how can we know this? A computer scientist at Cornell University, Jon Kleinberg, and his colleagues examined more than one million bond court cases in an effort to compare the efficacy of algorithmic predictions to the decisions of actual human judges. What they found would reinforce anyone's nervousness about the fallibility of human judgment. If decisions were made using algorithmic predictions, the level of crime by released defendants could be reduced by 25 percent without having to jail any additional people. Moreover, to maintain the current level of crime committed by released defendants, 42 percent fewer people could be jailed. In other words, we could massively improve the human welfare of some people by not sending them to jail at the bail stage of a criminal proceeding, without any additional risk to social safety. It's a win-win proposition that falls heavily in favor of algorithmic rather than human decision-making. Kleinberg and his colleagues argue that judges tend to release many of the people whom the algorithm identifies as especially high risk and the strictest judges tend to keep people in jail regardless of their level of risk. Their best guess is that with 12 million people arrested every year in the United States, the jail population could be reduced by several hundred thousand if an algorithmic tool were deployed. And it helps that it's incredibly cheap to implement. Like magic, it simply requires good administrative data and judicial record keeping and some statistical analysis! And, unlike a judge, it never tires, working in the middle of the night as well as over a morning cup of coffee. In fact, the case against human decision-making in criminal justice keeps getting stronger. One recent study showed that judges are much less likely to find in favor of the applicant in immigration adjudication when it's hot outside. All this can be added to more general evidence about how human decision-making can depend on morally ir-relevant things, such as whether your favorite football team won over the weekend (it makes you more likely to elect an incumbent!).

However, an idea that seemed unimpeachable in theory ran into significant obstacles. A number of leading civil liberties groups pulled their support for the legislation at the last minute over concerns that the reforms would perpetuate rather than diminish discrimination. The ACLU expressed concern that the proposed bill "is not the model for pretrial justice and racial equity," while the leader of a local advocacy or-ganization, Silicon Valley De-Bug, slammed the legislators for taking "our rallying cry of ending money bail and [using] it against us to further threaten and criminalize and jail our loved ones." In an unlikely alliance, some civil rights activists joined hands with the state's three thousand bail bondsmen to rally opposition to the new legislation. A coalition called Californians Against the Reckless Bail Scheme was formed and gathered more than 575,000 signatures in only seventy days to put a proposition onto the ballot in November 2020. Amid an onslaught of advertising, with advocates bemoaning the unfairness of cash bail and critics raising the specter of racial bias in algorithms, citizens went to the polls and decisively rejected the new law. Cash bail will now remain in place and algorithmic risk scores are on ice, at least in California.

The critics were onto something. Though deciding pretrial detention via algorithmic risk scores may be *better* according to some criteria, there is a lot of debate about whether it is *fairer*. A highly publicized ProPublica investigation of pretrial detention in Broward County, Florida, concluded that the algorithm used in that context was "remarkably unreliable"—little better than a coin flip in most cases. The algorithm was used to predict whether defendants would be likely to commit another crime—known as recidivating—if they were released from jail prior to their trial. On the surface, the algorithm appeared to treat black and white defendants somewhat similarly, correctly predicting if a defendant would recidivate 59 percent of the time for white defendants and 63 percent of the time for black defendants. If anything, one might conclude that the algorithm was biased against white defendants. But a deeper and more disturbing investigation of the algorithm's predictions uncovered evidence of significant racial disparities against blacks. Black defendants who would not have committed another crime were mislabeled as future criminals at a rate that was almost double that of white defendants (45 percent for blacks versus 23 percent for whites). Furthermore, white defendants who would have committed another crime were mislabeled as low risk 70 percent more often than black defendants. The results of the ProPublica investigation led to a great deal of scholarly debate, including about whether ProPublica had employed appropriate statistical measures, using a relevant definition of fairness (which has been disputed by Northpointe and others), or had made claims that were too strong in the face of other mitigating evidence. The debate continues, but the narrative of racially biased algorithmic decision-making has nevertheless taken hold.

Subsequent studies have reinforced the concern. In Kentucky, before the systematic introduction of algorithmic decision-making, white and black defendants were offered no-bail release at roughly the same rate. However, after Kentucky's legislature passed a law in 2011 that required judges to consult an algorithm in making their decisions, those judges began offering no-bail release to white defendants far more often than to black ones. Results such as these focus the mind on the details: what kinds of data are used to train algorithmic tools, how the models are constructed, how judges interpret the results, and so on. These concerns are now deemed significant enough that the Wisconsin Supreme Court mandated, in response to the Eric Loomis case, that warning labels be attached to algorithmic risk scores, alerting judges to certain "limitations and cautions."

California's bail reform left many of the details to be figured out on the back end. It would have been up to each of the state's fifty-eight counties to develop its own algorithmic decision-making tools. Each county, under the supervision of the state Supreme Court, would have decided what data to use, what model to build, and how to introduce risk scores into the judicial process. The county could develop an algorithm itself or procure an algorithmic tool from a company. The new approach would also not have been fully evaluated until 2023, four years after implementation, with the details of the auditing to be worked out later. Though the results of the 2020 election may have rendered these issues moot in California for the time being, similar challenges will need to be addressed in other domains as the use of algorithmic decision-making continues to roll forward.

Indeed, most of us will not face algorithmic decision-making in the criminal justice system, though we will grapple with algorithms in many other aspects of our daily lives. But those who will face it are among society's most vulnerable—and often the victims of historical injustice and systemic inequalities. Cathy O'Neil, a former mathematics professor turned data scientist, emphasizes this in her seminal book *Weapons of Math Destruction*, writing that algorithmic decision-making models "tended to punish the poor and the oppressed in our society, while making the rich richer." She recounts this phenomenon in the criminal justice system as well as many other domains such as credit scoring, college admissions, and employment decisions. Ruha Benjamin, a professor at Princeton, famously referred to this dynamic as the "New Jim Code," emphasizing how algorithmic models exacerbate existing racial hierarchies. If we are to embrace a future with more automated decisions, it needs to be one in which all of us have confidence. How can we achieve the promise of algorithmic decision-making while protecting society's commitment to fairness?

A NEW ERA OF ALGORITHMIC ACCOUNTABILITY

A first step might be to make sure we get the tools right. The first key decision the designer of an algorithmic model faces is what outcome to predict. In the case of Amazon, it was up to Beth Galetti and her team to define what it meant to be a "successful" employee at the company. It is not at all obvious how to define success in this context. For some job categories, one might measure success in terms of the efficiency or quality of the computer code an employee writes, but this wouldn't make sense if a person was being hired as a product manager or senior executive. One could note whether employees receive high performance reviews and get promoted, but these may be only loosely related to actual job performance and not to the interviewer's assessment of a candidate. Laszlo Bock, former senior vice president of people operations at Google, recalled an internal company study that looked at tens of thousands of interviews, established who had conducted each one and the score the applicant had been given, and finally, how that applicant had ultimately performed. "We found zero relationship," he said.

In the public sector, it's even less clear how to define the optimal outcome. Let's return to the criminal justice system. Should an algorithm be optimized to reduce the likelihood of individuals' not showing up to their court date? To reduce the likelihood of individuals committing a crime before trial? Or to minimize the likelihood of individuals ever committing a crime in the future? Optimizing for these distinct scenarios will have significant ramifications for who is denied bail. If California were to look to each of the counties to develop its own algorithm, it would almost guarantee unequal treatment for defendants across counties. And little guidance was provided for how those decisions should be made— whether by court authorities, elected officials, or even citizens themselves.

The California story points the way toward the first element of a good tool: it must have a clear, well-measured outcome that people accept as a legitimate objective.

Once an outcome is selected, we need to know that the model is accurate and valid. But in the real world, predictive accuracy isn't enough. We need to know how accurate the model is as compared to the best available alternative. In the case of Amazon, we would want to know whether the algorithm does a better job at identifying talent than the traditional interview process. In the criminal justice system, the corollary is the ability of judges to identify defendants who are at the greatest risk of jumping bail or committing another crime before trial.

A model is considered valid if its predictions map reasonably well onto the outcomes we actually observe in the world. The challenge is that what is predictively accurate in one context has no guarantee of traveling well to other contexts. If one is trying to develop a model to predict the likelihood of success for a potential hire at an Amazon fulfillment center, it may not make sense to use a model trained on data from workers at Amazon headquarters. If one is trying to develop a model to predict the likelihood of a defendant committing a crime while out on bail in the United States in 2019, it may not make sense to use a model constructed on data from Sweden or even from a different subset of Americans, such as individuals over the age of forty-five living in states that border Canada. Another factor is imperfect data quality. This is a really difficult issue in the context of predicting recidivism because we cannot know, for any given individual, whether he actually committed a crime; all we know is that the individual was arrested for or convicted of committing a crime. So the second requirement of a good tool is that it must be accurate and valid.

Getting the tool right also means being attentive to the risk of bias. In practice, the issue of bias cannot be easily separated from accuracy and validity: bias in the predictions can be the result of imperfect data quality, poor proxy variables, issues with the sample, and other factors. For example, in the Amazon case, if prior promotions favored employees who played well in a male-dominated environment, one might observe a lower rate of interviews for women even if the model did not explicitly take gender into account. Likewise, in the criminal justice context, variables that measure the characteristics of neighborhoods, such as income and whether one's friends have been arrested, are closely associated with race and thus generate systematically different outcomes across racial groups even if race is not explicitly included in the model. The only realistic strategy for engineers addressing bias is to systematically measure and take steps to mitigate bias and to be explicit about what conception of fairness they are optimizing for.

THE HUMAN ELEMENT IN ALGORITHMIC DECISIONS

Getting the tools right is not enough; we must also pay attention to how they interact with human beings—preferably before we deploy them at scale, as California proposed to do. Because if human beings ignore the recommendations of algorithms or are able to game automated systems, we won't achieve the amazing outcomes that computer scientists predict we will when they are tinkering in the lab.

The accuracy of a tool is not necessarily correlated with its efficacy in the world because the ultimate decision-making authority in most contexts still rests with human beings. So to understand efficacy, we need to observe how human beings interact with algorithmic predictions. One strategy would be to run some sort of experiment, ideally a randomized controlled trial; for example, identify a set of counties and randomly assign half to make bail decisions via algorithmic predictions (treatment group) and the other half to rely on judges' decisions (control group). With a sufficient number of counties, you could then track the likelihood that individuals would show up to court and the rate at which defendants commit crimes in the two groups. If you observe higher court appearances and lower crime rates by defendants in the treatment group, you'd have some compelling evidence of the efficacy of the tool.

One might naturally think that California legislators would have wanted such an experiment to be run before adopting algorithmic decision-making tools statewide, but in fact they moved forward without one. Although such experiments are ongoing, at the time of this writing none has yet been completed. So algorithmic decision-making, at least in the judicial system, has grown by leaps and bounds in the absence of systematic evidence of its efficacy. The decision to use algorithms, especially when dressed up in the language of artificial intelligence, is often treated as a magic potion by its most naive enthusiasts. This is nothing more than wishful thinking.

Even in the absence of an experiment, we can get a handle on one obvious step where things might go awry: the point at which a human decision maker receives an algorithmic recommendation and must decide what to do. Some observers have been concerned that humans tend to unquestioningly accept the accuracy of an automated decision, known as automation bias. The reality, however, is that human beings often ignore recommended decisions, doing so in ways that systematically harm some groups and not others. Recall Kentucky's move to mandate the incorporation of algorithmic risk scores into judicial decision-making, which led to higher rates of pretrial release for white defendants than black defendants. This could be a function of judges following the recommended approach at equal rates for whites and blacks and the algorithmic risk scores accurately capturing different levels of risk. But another study found that judges are more likely to override the recommended judgment for black defendants than for white defendants, even when applicants share an assessment of low risk. Worryingly, these effects were especially concentrated in whiter counties. This suggests not only that judges didn't trust the system (in fact, they rejected the recommendations two-thirds of the time) but that they interacted with the recommendations in ways that reinforced some of their own biases.

Even if judges fully accept automated decision-making tools, in practice they might not be as effective in the real world as predicted in the lab. Perhaps the main reason is that human beings respond in both predictable and unpredictable ways to changes in their environment. Predictably, if an automated algorithm determines whether someone's résumé is flagged for an interview, those who know how the algorithm works might game the system by offering advice to future applicants, perhaps even for a price. Amazon staffers who want their friends and colleagues to be hired by the company might develop ways of working around the automated résumé-screening tool, perhaps by pushing their own favored candidates more strongly through personal networks, thus rendering the tool less efficacious overall. Likewise, in the criminal justice system, there might be incentives for defendants to report (or hide) information that might enable them to take advantage of the automated decision-making tool. A lawyer might advise defendants about what to share and not to share with the police when they are picked up so as to avoid pretrial detention.

In a totally different context, the US Air Force Academy experimented with algorithmically matching first-year students to peer study groups. It made the move because a machine learning algorithm detected from historical data that low-performing students did better when matched with high-performing students in a study group. However, when the information was acted upon, cliques quickly developed: high- and low-performing cadets didn't want to hang out with each other, and the low-performing students were left behind. The algorithm seemed to miss the fact that a strategic effort to optimize group design would expose differences in a way that undermined the value of mixing groups altogether.

These examples underscore the importance of figuring out whether a tool works before deploying it at scale, a sort of "do no harm" principle for algorithmic decision-making. Yet even though systematic study of how humans treat algorithmic recommendations is in its infancy, it hasn't slowed down the enthusiasts at all.

HOW TO GOVERN ALGORITHMS

Given that fairness is something we must mutually agree upon, how should we govern the use of algorithmic decision-making tools, whether in the private or public sector? Right now, most people don't even know when an automated tool is being used to make decisions that impact their lives. This matters because a natural response to a decision-making process that is not understandable is to see it as unfair. And any perception of unfairness, especially in regard to the decisions made by public sector institutions, is a significant threat to legitimacy.

Democracies around the world are introducing measures to govern how automated decision-making algorithms are used. The Canadian government is in the forefront, having adopted a nationwide algorithmic impact assessment. Similar provisions are included in new European directives. And in the United States, a few years ago, a popular member of the New York City Council decided to tackle these issues head-on. James Vacca, who was serving his last term in government, had grown increasingly concerned that too many public decisions were being made on the basis of opaque algorithmic systems. For example, when he complained to the city that there weren't enough police officers in several precincts in his district, the New York Police Department told him it had a formula to determine how manpower should be allocated across the city. And when a parent came to him with concerns that her child had been assigned to her sixth-choice public high school, Vacca could do nothing more than reference the Department of Education's mysterious school-assignment algorithm.

On the one hand, Vacca was enthusiastic that New York was at the forefront of using data to improve the delivery of city services. But he was also deeply troubled that those on the receiving end of these decisions were not even aware of how they were being made or how they might challenge them if they seemed wrong. Vacca proposed legislation to mandate that city agencies publish the source code of all algorithms and allow members of the public to self-test them by submitting their own data and getting the results. During a committee hearing in 2017, he framed his bill as a key to democracy in the digital age: "In our city it is not always clear when and why agencies deploy algorithms, and when they do, it is often unclear what assumptions they are based upon and what data they even consider. . . . When government institutions utilize obscure algorithms, our principles of democratic accountability are undermined."

Vacca's proposal was the first of its kind anywhere in the United States. As the proposal made its way through committee, the complexity of the task was made apparent. How much transparency around algorithms was the right amount? How could the city protect against the risk that algorithms would be gamed if the source code were available? Who would be responsible for testing the source code and evaluating its quality? Strikingly, he discovered that senior officials in the Mayor's Office of Data Analytics could not provide answers to questions that he and his colleagues were raising about particular algorithms. The reality was that even elected officials were unable to get information about algorithms that were making critical decisions about New Yorkers' lives.

Ultimately, Vacca's mandate was replaced with a watered-down task force to examine how automated decisions were being used. After two years of work, the task force issued its final recommendations. A coalition of local nonprofit organizations and civil liberties groups released its own "shadow report" alongside the city's findings, pushing the city to go even further and arguing that New York's leadership on algorithmic accountability will have an "outsized influence on current global policy debates."

Together, these reports point to three key ingredients in the governance of algorithmic decision-making. The first is transparency. Think of this as the equivalent of disclosure requirements for food packaging. People should know when automated decision-making systems are being used to make determinations that directly affect them. In addition to knowing *that* an algorithm is at work, people should understand *how* the algorithmic predictions are used and have the ability to access relevant policy and technical information that reveals how the algorithm was designed, how it works in practice, and how it has been evaluated in terms of its impacts.

We cannot be naive about the value of disclosure on its own, however. For many algorithms, the code itself will not be all that revealing when it comes to particular determinations, and it is unrealistic to rely on members of the public—or even civil liberties groups and government watchdogs—to fully police the details of automated tools. But as Jon Kleinberg and his colleagues have argued, with appropriate disclosure, potential discrimination in algorithmic systems will be rendered even more visible than discriminatory behaviors by human beings, perhaps enabling greater progress toward fairness and justice.

So the second key ingredient is auditability. Whenever possible, algorithms should be independently tested and validated, with the results made publicly available, including with explicit checks on bias. In the private sector, a commitment to auditability is much harder to achieve; the algorithms themselves may be proprietary, and unless there are allegations of discriminatory effects, companies are likely under no legal obligation to make their decision-making tools available for public auditing. However, when it comes to public sector institutions, the decision to empower, enable, and resource independent oversight bodies is a no-brainer. One cannot simply rely on citizens to do the checking, and the process in New York made clear that neither city officials nor elected council members are well positioned to do this work. Expertise is needed, and organizing and deploying this expertise will be one of the costs associated with harnessing the efficiency gains of these systems.

The final ingredient is a commitment to due process. Individuals and groups should know how they may challenge decisions informed by an automated tool as well as the procedure for doing so and the response time after a request or complaint is lodged. Again, this is particularly important for public institutions such as courthouses, schools, police stations, welfare offices, and the tax collector. But it is natural to think that companies using these tools will need to develop due process mechanisms as well (for example, in the case of loan determinations or the settings of insurance rates). Such mechanisms will be especially important in addressing perceived concerns about the disproportionate impact of automated decision-making tools on particular demographic groups, including women and communities of color.

Though there is broad agreement that transparency, auditability, and due process are the principles needed to govern algorithmic systems, it shouldn't surprise anyone that the devil is in the details. And working out the details isn't something that most of us will engage in. But it is something we should look to our elected officials to do, well in advance of the systems' mass deployment in new and sensitive domains, such as the criminal justice system. Although most of Vacca's constituents probably had no idea what he was up to, his work shows how effectively we can use our traditional democratic institutions to lock in the rules that fairness demands in this new age of algorithms.

Indeed, other members of the New York City Council continue to push forward where Vacca left off. One effort introduced in 2020 by Laurie Cumbo specifically focuses on automated tools for hiring decisions. The bill requires employers who use such tools to notify job candidates and also inform applicants about how the tools were used to screen their applications. Moreover, the bill requires that such tools be subject to annual audits for bias. So far, the bill has received mixed reactions. Some call it a good first step and a way to get more of the public engaged in conversations about algorithmic decision-making tools. Others worry that such legislation doesn't provide enough details with respect to auditing requirements and could allow a vendor to sell software under the imprimatur of its being "fair" even when the tools have not been rigorously evaluated. The debate continues, and it won't end with this bill.

OPENING THE "BLACK BOX"

Some people question whether algorithmic transparency is even possible. Many of the "deep learning" models that are built today include many millions of parameters. Understanding how every one of them was affecting the model's decision-making would be too complicated to figure out. But the engineers that built the model based its architecture on their understanding of the problem they were trying to solve. We can require the engineers to produce a high-level description of the model that would lay out their understanding of what it was supposed to encode. This could be augmented by various types of sensitivity analyses, tested on different gender, racial, and socioeconomic groups to see if they are treated similarly along dimensions that should not impact the decision-making process. Such auditing will give us information about the reliability of the algorithm across different measures of fairness.

Researchers have suggested ideas such as building simpler models that are easier for humans to interpret. We might choose to give up some small amount of accuracy to obtain a more readily understandable model, thereby better balancing the value of predictive accuracy against the value of transparency.

But remember that the process of human decision-making is also opaque. We can't inspect what's going on in someone's head when he or she makes a decision. That doesn't stop us from allowing humans to make them. The goal is not to require complete transparency or reliability before we can trust an algorithm to make a decision. Rather, we need to understand enough about the decision-making process and assessments that have been made of the model results to believe that it produces better, fairer decisions than the alternative does.

Of course, transparency at an aggregate level is not enough. If an algorithmic model decides that you should not be granted bail and has been measured to be "fair" across genders and races *in the aggregate*, you should still have a right to challenge the decision *as an individual*. The right of an individual to appeal a decision is a time-honored process, especially in criminal justice, and should be available even if an algorithm makes the decision. One way such an appeals process could be implemented is to have a human responsible for justifying the answer given by the algorithm or overturning the decision if it is found to lack merit. This creates an incentive for the model creator to make it as transparent and understandable as possible. It also opens the door to creating legal liability for the producers of automated decision makers, giving them an even greater incentive to conduct thorough audits of their models and put even more effort into making them more transparent. The onus for building fair decision-making systems shouldn't fall on the individuals about whom decisions are being made. Rather, regulation needs to compel the developers of such systems to make them more understandable.

● ■ ● ■ ●

You might now be tempted to think that requirements for transparency and auditing are simply unrealistic for algorithmic decision-making systems. But let's return to the example of Amazon's résumé screener. Here was an algorithm that could have been broadly deployed to screen millions of résumés, ultimately leading to the perpetuation of gender bias in hiring at one of the largest companies in the world. Instead, internal auditing showed that the model did in fact have a gender bias. That was followed by attempts to understand and correct the flaw, and when it was clear that it couldn't be fixed, the plans to release the algorithm were shelved.

The résumé-screening model was a failure, but the process that led to its analysis, attempted repair, and eventual abandonment was not. In fact, that's just how algorithmic decision-making systems should be handled. Moreover, the auditing process for the algorithm helped uncover deeper problems related to the role of gender in historical hiring practices, helping spotlight the bias in the data used to train the system in the first place. The truly frightening alternative would have been for the model to be developed, never audited, and simply deployed without recourse. It gives one pause to consider the many situations in which just such algorithmic decision-making systems have been deployed by overeager engineers who never stopped to check their algorithmic biases.

We can't leave it up to the goodwill of companies and government agencies to do the hard work of auditing and refining their algorithms while shielded from public view. Instead, we need to create an expectation that they will do so, transparently, as part of a new era of algorithmic accountability. Many decisions that impact our lives will depend on it.

CHAPTER 5

WHAT'S YOUR PRIVACY WORTH?

Most people know Taylor Swift for her Grammy Award–winning pop music and the countless romantic breakups that have fueled her hit singles. Few people realize that she faces a constant stream of threats from stalkers who have, in her own words, "showed up at my house, showed up at my mom's house, [and] threatened to either kill me, kidnap me, or marry me." One man broke into Swift's town house in New York in 2018, used her shower, and then took a nap in her bed. Immediately after serving time in prison for that invasion, he returned to her Tribeca home while on probation and smashed a window to gain entry, before being arrested by the police. He has tried to gain entry to Swift's home three times.

Enter ISM Connect, a firm that uses the latest in facial recognition technology to "simultaneously enhance security, advertise and collect demographic data for brands." Swift contracted with ISM Connect during her 2018 tour to protect herself from stalkers. Using its trademarked FanGuard technology, ISM installed cameras behind kiosks marked as "selfie stations" at Swift's concerts. The selfie stations enticed concertgoers with Swift memorabilia and behind-the-scenes video footage. As fans interacted with the content, hidden cameras captured images of their faces. According to a security contractor who spoke with *Rolling Stone*, the data were then sent to a central command team in Nashville to check against a database of known Swift stalkers. Not losing an opportunity to harness the data for additional purposes, ISM also uses its smart screens to capture demographic information and metrics that will help brand promoters direct their marketing efforts.

Consider what's at stake with this new technology. Fans are tricked into presenting a frontal view of their faces to a camera, and the images are recorded, stored, and transmitted to a security and marketing company. The company emphasizes that signs inform crowds that "they might be filmed," but no consent is sought and concertgoers have no control over how their images are used. Add to this the fact that Taylor Swift concerts attract many children and teenagers, whose facial images are being captured as well.

The idea of protecting Taylor Swift from known stalkers makes sense. But greater safety is something we all want, not just celebrities who already have their own private security teams. Cities around the world are using a mix of linked cameras, aerial surveillance, and facial recognition technologies to make the apprehension of criminals more likely and deter future criminality. The city of Baltimore is a case in point. With its high crime rate and a police force under scrutiny for bias, excessive use of force, and corruption, a group of community residents has been lobbying for the adoption of an aerial surveillance system. The idea behind this technology is simple: a plane would hover over the city, capture a constant stream of images, and stitch them together for a second-by-second portrait of activity on the ground. This kind of imagery would enable a revolution in policing. Police officers would no longer need to rely only on the fragmentary evidence at a crime scene to uncover potential suspects, check their alibis, and build a case before making an arrest. Instead, the images could be used to pinpoint the moment of the crime and track the people and vehicles at the scene backward and forward in time, using aerial data and street cameras to identify and locate potential suspects who were in the area.

Of course, facial recognition technology can suffer from some of the same sorts of algorithmic biases that we saw in chapter 4. Joy Buolamwini, the founder of the Algorithmic Justice League and a researcher at MIT, and her coauthor, Timnit Gebru, a cofounder of Black in AI and a leading researcher in ethics and AI, documented significant gender and racial discrepancies in the performance of facial recognition systems from Microsoft, IBM, and the Chinese platform Face++. Their work shows that such systems perform worse on females and people with darker skin, with errors compounded for dark-skinned females. After they communicated the results to the companies, the companies took steps to try to make their systems perform more equitably across groups, but some differences still persist. Given the interest by law enforcement officials in deploying facial recognition systems to identify individuals engaging in criminal activity, errors made by such systems can not only create legal entanglements for otherwise innocent people but also exacerbate racial inequality if they are more likely to be made for darker-skinned individuals.

Still, there are benefits of surveillance technologies we might consider beyond crime prevention. With the outbreak of the novel coronavirus early in 2020, a chorus of technologists called for rapid innovations in digital disease surveillance. Why stick to an outmoded model of manual contact tracing, with health care workers monitoring the movement of infected individuals, when digital technologies could enable large-scale tracking via GPS devices, cell phone towers, Bluetooth connections, internet searches, and commercial transactions? The tech optimists pointed to South Korea's success against the coronavirus, where widespread testing combined with a data-driven approach to tracking movement led to rapid declines in the COVID-19 infection rate. In pursuit of a fully modern approach to contact tracing, Google and Apple announced an unprecedented partnership to develop a contact-tracing app that would use low-level Bluetooth signals to alert anyone whose mobile device had come near an infected person in the past two weeks. As the partnership got off the ground, however, the obvious questions about data access, ownership, and protection rose to the fore.

What the British philosopher and social reformer Jeremy Bentham proposed in the eighteenth century as a tool of desirable social control— an omnipresent surveillance of prisoners, or "panopticon"—is no longer just a philosopher's fantasy. But nobody expects, much less desires, to be tracked from moment to moment, with the intricate details of our lives pieced together and made permanently reviewable by companies or governments. Just because a sign notifies us that cameras are operating nearby, it doesn't mean we've given our consent to allow data about our lives to be collected or that we fully understand how such data might be used.

The case of surveillance technologies reveals a wider set of tensions between data privacy and personal safety, national security, research and innovation, and convenience. Over the past decade, most people have come to accept a world in which we routinely sign off on terms of service for every app and digital product without so much as reading those terms, thereby giving tech companies enormous leeway to hoover up the data exhaust stream from our digital activities. Tech companies are extractive industries just like oil companies. The process is called *data mining* for a reason.

If we're honest, however, we have to admit that the aggregation and analysis of data have given us an incredible array of digital tools and products: warning us about road traffic and suggesting quicker routes, predicting what we're searching for after only a few keystrokes, and recommending with surprising accuracy what we might want to watch or listen to next. We gain the convenience of these free services in exchange for our data. Google search is available to anyone with internet access, no fee involved. Your Apple Watch wants not only to record your heart rate and share the data with your doctor but also to stream it into a data ocean for researchers to learn more about human health.

So we're faced with a dilemma: Information about where we go, whom we see, what we read and watch, with whom we communicate, how long we sleep, and biometric data points such as our faceprints, fingerprints, and heart rate—these are things we think of as private. Yet that same information, in the hands of others, can serve as the basis for an amazing array of possibilities, including the convenience of personalized services, medical innovations that might save lives in the future, and the protection of citizens from domestic and foreign threats. These are great benefits, but it appears that they can't all be realized without significantly threatening our privacy. The data can also be used in unanticipated ways, such as when the company Clearview AI scraped billions of images from the internet to build a facial recognition system that can be used to identify virtually anyone (most likely including you) in real time. A prospective investor in the company, billionaire John Catsimatidis, used the app to identify a man he happened to see on a date with his adult daughter. It was also used by law enforcement to identify rioters who took part in the assault on the US Capitol in January 2021.

Governments and citizens have been in a constant tug-of-war over data since the very inception of what we think of today as a "state." What's new is that we've willingly handed over to private companies the permission to collect our personal data almost without constraint, creating an entire political economy aptly labeled by Harvard professor Shoshana Zuboff "surveillance capitalism." By contrast, democratic governments are far more protective of personal privacy because they value individual liberty and therefore impose limitations on their own ability to collect and use data. Taylor Swift didn't ask the police to set up a facial recognition system for fans at her concerts; she asked a private company to do it because there were far fewer restrictions on what kind of data it could collect and what it could do with it.

This new reality demands that we ask some hard questions about privacy: Are we entitled to it? Where and when must it be preserved? Under what conditions are we willing to trade it for safety, innovation, and convenience? Should we feel differently about sacrificing our privacy to a government, with its mission to protect the public interest, as compared to private companies that are maximizing shareholder returns?

THE WILD WEST OF DATA COLLECTION

Controversy around privacy rights in the digital age first received widespread press coverage in the United States in the mid-1990s, when the White House announced the Clipper Chip. Responding to the growing threat of encryption as a technology that could prevent law enforcement from listening in on phone conversations that it deemed might threaten national security, the government wanted to create a means of preventing its wiretapping capabilities from "going dark." Enter the Clipper Chip, a government-developed technology that could be embedded into consumer electronics to allow for encrypted communication while still providing government with a "back door" for law enforcement to listen in on private communications under court order. The response to the Clipper Chip was swift and overwhelming, with an array of individuals and organizations in industry, civil society, and academia criticizing the program for having been developed in secret, lacking social and technical safeguards, and likely not providing the level of security or privacy that the government or consumers would hope for. Some odd bedfellows, including the American Civil Liberties Union and the conservative talk show host Rush Limbaugh, found themselves on the same side of the debate. Faced with mounting criticism, the Clinton administration abandoned the Clipper Chip within a few short years.

In the wake of the 9/11 terrorist attacks, the US government had a renewed sense of urgency about monitoring telecommunications. The justification now wasn't just law enforcement but the more widely held concern that future terrorist attacks needed to be stopped at all costs. Through the USA PATRIOT Act, the government enacted sweeping programs to collect phone records and other digital communications, touching nearly everyone in the United States, no judicial authorization required. Ultimately revealed by documents leaked in 2013 by Edward Snowden, a contractor for the National Security Agency, these programs led to a massive global outcry against the government's privacy infringement overreach. The programs in question were eventually reviewed and a number of reforms adopted to curtail some of the more egregious forms of data collection, but the damage had been done: the public had become even more sensitized to the danger of government violation of basic expectations about personal privacy.

Despite the public concern of government surveillance programs and calls for greater regulation, information collection by private companies received little scrutiny by comparison. Indeed, until quite recently, the private sector has had scant regulation with respect to the information it can collect and process from the users of its products or systems. Companies such as Google and Facebook have been collecting troves of user interaction information—search queries, web pages visited through their platforms, friend connections, and "likes," among others—for the purpose of honing their ad-targeting capabilities. The ability to optimize ad targeting with surgical precision quickly became the engine of the enormous revenues at these companies. Google alone generated more than $130 *billion* in advertising revenue in 2019 as a result of employing large-scale machine learning algorithms to take its ad-targeting capabilities to unparalleled levels.

By contrast, companies without the same ad-targeting savvy are relegated to the pantheon of "also-rans" in the tech sectors. Yahoo!, one of the early pioneers of the internet, which turned down an offer to acquire Google's search technology when Larry Page and Sergey Brin were still graduate students, was sold to Verizon in 2017. Its sale price: $4.5 billion, or less than one-twentieth of the $95 billion in ad revenues Google generated that same year.

Why is there so much uproar about government surveillance and so little when it comes to the private sector? It's puzzling, as both are exploiting reams of personal information. You might think we would give the government greater latitude since it is responsible for guaranteeing our security, especially from external threats such as foreign terrorism. Moreover, much of government surveillance goes unseen. If a terrorist plot is prevented and you never hear about it, it's hard to appreciate the fact that some infringement of your privacy may have been necessary to stop it.

The ways in which private companies are harvesting our data are far more visible. In addition to monitoring our communications, tech companies tend to gather data on our interactions, which reveal our interests—what we search for, click on, and like. In return, these companies provide direct, tangible benefits, frequently at no financial cost, such as email accounts, access to information, the ability to stay more easily connected with friends, and applications to share even more information.

As the scope of these platforms grows, we find ourselves sharing more personal information through instant messages, emails, images, videos, and voice commands. We're no longer using the platforms just for getting information and sending a quick message or two. More and more, the platforms are becoming our dominant means of communication, the way in which we produce, distribute, and consume information among our family, friends, and strangers.

Realizing that so much of our contact with others is directly enabled by technology, we might rightly ask what sorts of safeguards exist for the information we so eagerly provide. The Federal Trade Commission, which touts itself as "protecting America's consumers," promotes a doctrine known as "Notice and Choice" (sometimes also referred to as "Notice and Consent"). The basic idea is that companies that collect data must provide *notice* to their consumers. That is, they should tell potential users what information may be collected about them, how that information may be used, and what the company's policies are regarding storing that information over time. Such notices are often embedded in a product's impenetrable "Terms of Service" or in a "Privacy Policy." Users then have the *choice* to consent and use the product under those terms or to decline, which generally means they won't be able to access the product or install the app.

Of course, you're all too familiar with this. You know the long document of legal gobbledygook you're asked to read and approve when you want to sign up for an account on a website or install an app—well, you were just given *notice*. And if you don't actually try to read and decipher the whole thing and instead happily—or grudgingly—click "Accept," you've made your choice. Of course, you aren't alone. In a 2017 survey of mobile device users, Deloitte reported that the vast majority of consumers using mobile apps "willingly accept legal terms and conditions without reading them."

The first question is: What are we giving up? Here's an excerpt from Facebook's Terms of Service as of spring 2021:

> *Specifically, when you share, post, or upload content that is covered*
> *by intellectual property rights on or in connection with our Products,*

you grant us a non-exclusive, transferable, sub-licensable, royalty-free, and worldwide license to host, use, distribute, modify, run, copy, publicly perform or display, translate, and create derivative works of your content (consistent with your privacy and application settings).

When you upload a picture to Facebook, it can sublicense that picture to someone else, potentially after modifying it or creating a derivative work. Maybe a picture you posted of your family would make a great stock photo for a news story about dysfunctional families. Of course, that can be done only if it's consistent with your privacy and application settings—which are often just as likely to be ignored as the original terms of service are. This is not to say that this kind of photo reuse is actually done, simply to provide a cautious reminder that we may be giving up far more than we think when making our "choice" to use an application. In the Notice and Choice framework, the burden is squarely on consumers' shoulders to take responsibility for their privacy, even if it entails having to decipher pages of legalese, navigate baroque privacy settings—with the default almost always against privacy protection—or accept onerous terms of service lest one lose access to information or be left out of social communities.

At a panel discussion we hosted at Stanford in 2019, the shifting landscape of privacy concerns was on clear display. Jennifer Lynch, the surveillance litigation director at the Electronic Frontier Foundation (EFF) practically embraced Rick Ledgett, Jr., a former deputy director of the National Security Agency (NSA), in celebrating the government's accountability mechanisms in data-gathering as compared to the scant limits placed on private companies. In the era of the Clipper Chip, EFF and the NSA were far more likely to have opposite views of how government regulation should approach privacy concerns. Though EFF and the NSA still have their differences, they now share many of the same concerns about the policies of tech companies represented by fellow panelist Rob Sherman, the deputy chief privacy officer at Facebook. When asked how Facebook viewed user privacy concerns, he drew audible laughter from the audience when he stated the company's commitment to "putting user privacy first."

Facebook isn't the only example or even the worst. In some cases, decisions regarding user privacy are made by software engineers and product managers, who believe they are creating something in users' interest without realizing the broader implications of their work. They're optimizing for how *they* view interacting with the world, not necessarily thinking about the privacy concerns of a diverse user base. That myopic viewpoint was on full display in the 2010 launch of Google Buzz, which aimed to combine social networking and email. It was heralded as a way for users to share information across a variety of media. By default, Google Buzz made publicly available the list of people users emailed or chatted with most often. Why wouldn't someone want to connect more with everyone they knew? It seemed like a great way—at least for the people building the product—to help people stay more connected to the list of everyone they had been in contact with. Well, for one user, that list included her abusive ex-husband, who had the ability to access comments she'd posted to her boyfriend, including her current location and workplace. It's likely that she was one of the vast majority of people who didn't completely read or appreciate the terms of service before making the "choice" to check "Accept." Or maybe it wasn't clear to what extent her information would be made available or to whom. Whatever the case, it's yet another example that the currently accepted framework of Notice and Choice fails to provide the kind of privacy protection we need and should demand in the online world.

A DIGITAL PANOPTICON?

In a public lobby of University College London, you will find a most peculiar sight. The embalmed corpse of Jeremy Bentham, clad in one of his favorite black suits, sits on a chair accompanied by his cane. Encased in a glass-fronted wooden box with the label AUTO-ICON, he appears to gaze out upon the hundreds of people who pass him every day. The arrangement is not due to some macabre wish to honor the eighteenth-century British philosopher; it is a reflection of Bentham's own wishes, expressed in his 1832 will. In addition to detailing the transformation of his body into an Auto-Icon, Bentham's will specified that on occasions when his friends and disciples gathered at the university to discuss utilitarianism, the box should from time to time be taken to the room and stationed as if to take part in the discussion. Philosophers can be strange, even in death.

Bentham is widely regarded as the founder of utilitarianism, a philosophy that espouses achieving the greatest good for the greatest number of people. In essence, utilitarianism renders ethics as a system of moral mathematics. Not just a philosopher scribbling out the abstractions of utilitarianism, Bentham was also an important social reformer. He deployed the ideas of utilitarianism to justify a wide range of progressive policy changes. He engaged in the policy debates of his day and bequeathed to future generations not only a moral framework that would come to be adopted by economists in the twentieth century and engineers in the twenty-first century but a set of policy proposals, many of them quite radical, meant to deliver enormous benefits to society.

Perhaps the most famous of such proposals is his idea of a panopticon. The panopticon refers to the architectural design of a prison where a watchtower sits in the middle of a circular edifice of prison cells. The tower shines bright light from the center outward so that the watchman is able to monitor everyone in the cells. Why call the design a panopticon? It was Bentham's invented term for an all (pan)-seeing (optic) mechanism, a novel construction that would allow a watchman to observe the prison's occupants without their knowing whether they are being watched.

Prisons were dangerous, dirty places. Bentham thought his proposal to construct prisons as panopticons would be a great step forward, allowing for cleaner, safer, and more efficient systems of incarceration. Fewer guards would be required while achieving an improvement in security because of easy surveillance, so costs would be lower and outcomes better. All privacy for prisoners was sacrificed in the name of improved security. The preface to his short treatise "Panopticon," in which he introduced the idea to the world, offered a catalogue of the social benefits to be gained from his clever "inspection-house": morals would be reformed, health preserved, industry invigorated, and public burdens lightened.

Bentham was clear-minded about why the panopticon could be so powerful: it was a revolutionary unidirectional form of psychological control over prisoners. He described it as a "new mode of obtaining power of mind over mind, in a quantity hitherto without example: and that, to a degree equally without example, secured by whoever chooses to have it so, against abuse." The better the surveillance, the better the control over prisoners, and at little risk to the guards. "It is obvious," he wrote, "that in all these instances the more constantly the persons to be inspected are under the eyes of the persons who should inspect them, the more perfectly will the purpose of the establishment have been attained."

Prison reformers in the twentieth century adopted Bentham's proposal and built a number of prisons based on his panopticon design. One of many, the last roundhouse Panopticon prison, F-House at the Stateville Correctional Center in Illinois, closed only in 2016 (and temporarily reopened in 2020 to serve as a quarantine center for inmates infected with COVID-19). But the concept continues to inspire the design of other prisons in the United States, including the Twin Towers Jail in Los Angeles, and in countries around the world, including France, the Netherlands, and Cuba.

Today, however, the idea of a panopticon summons to mind not progressive prison reform but dystopian social control and a surveillance state. The French philosopher Michel Foucault invoked Bentham's idea in the 1970s to describe the growth of modern techniques of surveillance in which inspections are routine; the gaze of the employer or the police or the border guard is frequent; and citizens need to register themselves, their activities, and their possessions—a car, a driver's license, permission to enter a school, a courthouse, or public park or to organize a protest, and so on—with multiple authorities. The growth of surveillance made its presence, paradoxically, more insidious and hidden. Modern conditions of surveillance amount, Foucault thought, to an omnipresent panopticon of unparalleled social control. We accustom ourselves to its operation, and we adapt and conform to its power. Surveillance could be permanent in effect, even if not actually present at any moment, because citizens need only internalize the belief that they could be, and likely are being, monitored. The upshot was a curtailing of our freedom by an invasion of our privacy.

All of which leads to the obvious question: Given modern technology, do we now live in a digital panopticon where nearly every last bit of privacy has been eroded? We know that we are observed a millionfold more than before the advent of digital technology and unimaginably more than would have been possible in Bentham's day. The prison panopticon seems positively quaint by modern standards.

As we've seen, the Wild West of data collection by tech companies goes far beyond what the pessimistic social critics of old could imagine. The automated collection of every click, like, and Web search is just the familiar tip of the iceberg of the digital surveillance of our behavior. There are the voice recognition services ("Alexa" and "Hey, Siri!") that collect what we say, the facial recognition (every digital photo service that tags you, your family, and friends) that analyzes our faces, the biometric fingerprinting (our retinas, our thumbprints, our gait) that enables swift passage through airport security or access to our digital devices, and the location tracking in our smartphones that helps us navigate when we drive or call an Uber or Lyft. There are digital card swipes that trace and track our entry to our workplace or home, along with video doorbells, digital thermostats, fitness trackers, and on and on. These wondrous technological achievements, most of which became ubiquitous only in the last decade, are all fueled by massive advances in computing power, machine learning, and the availability of data. It's a self-reinforcing loop, a virtuous or vicious cycle depending on your point of view: the collection of more data expands the capacity of powerful algorithms to make even more accurate predictions, which in turn leads to the development of new algorithmic models and new digital tools, which can be expanded for use in different domains for different purposes, leading to the collection of still more data.

Nothing captures the situation better than a cunning and playful project created recently by students at University College London. Deploying modern-day technology, they installed two webcams, one in the public lobby pointing toward Bentham's Auto-Icon and another atop the Auto-Icon pointing outward toward the people who pass by to gawk at his body encased in a box. The "Panopti-Cam" delivered a constant livestream to anyone in the world of people looking at Bentham and of the view of Bentham gazing at passersby. You could watch Bentham watching others omnipresently—all from the comfort of your home.

Thinking about a digital panopticon makes painfully clear what is lost in a world when everything is observed: we lose our privacy, and with that, we undermine our freedom, diminish the possibility of intimacy, and compromise our capacity to control what others know about us. As Bentham so clearly understood, surveillance is a means of social control, an opportunity for others to exercise power over us.

Sometimes it's obvious why the loss of privacy is bad. Just as we don't want strangers peering into our bedrooms, we have good reason to fear Alexa listening to us in the bedroom. We don't allow our health care providers to share our medical records with others, so why should we allow a fitness tracker to share our health data with others?

Yet sometimes it's not so obvious why a loss of privacy is worrisome. What's the big deal about privacy? you might wonder. If I'm not doing anything wrong, I don't need to hide anything from other people. Or you might think that privacy is about what happens in your home, not when you're out in public.

But the value of privacy is not about shielding others from seeing what we do illicitly or what we might be embarrassed to have others find out about. The reason we keep our medical records private is that we might be adversely affected if our employers, our medical insurance companies, or pharmaceutical companies knew everything about our health history.

The value of privacy extends far beyond the intimate quarters of the bedroom and our own dwelling. Privacy is important in public arenas, too. Consider a story told to us by one of our students. She said that her mother had won a prize for her work as a schoolteacher and was invited to the White House with other teachers for an award ceremony. Upon presenting her credentials at the White House gate, she had been denied entry. When she'd asked why, she learned that her face had come up in a database of people who had participated in public protests against President Trump, and protestors were not welcome.

She had done nothing wrong. She had nothing to hide. Her face had been recorded while she was behaving lawfully in public, and the records had been entered into a government database. It's easy to see that with surveillance comes a potential loss of essential liberties in a democratic society, such as freedom to protest and freedom of expression. It's no accident that the protestors in Hong Kong in 2019 took to wearing face masks. It wasn't to protect themselves from the coronavirus, which had yet to emerge. It was to conceal their faces from the omnipresent surveillance cameras that could allow China to record their activity and register them as enemies of the state.

At its heart, privacy is about the importance of control over oneself and information about oneself. Privacy protects the experience of intimacy, an essential part of any life, by shielding aspects of our relationship with loved ones that are not the business of others to know. Privacy allows us the ability to exercise other freedoms, such as protesting, without fear that we will pay a price. And privacy enables us to share information with some people, such as our doctors, and block it from being shared with others, such as our employers.

The importance of privacy has shifted over time. You won't find an explicit mention of privacy in the US Constitution, but it does appear in a few state constitutions and in a number of more recent international human rights documents. That's not by accident. The rise of social control through surveillance technologies is what has brought the value of privacy to the fore. Even when we acknowledge its importance, we are at different moments willing to sacrifice privacy in order to obtain other benefits such as security (think of Taylor Swift), convenience (think of the free apps on your smartphone), and innovation (think of personalized medicine). But none of this is to say that privacy has no value. And with the arrival of a digital panopticon, there has never been less weight placed on privacy.

Indicators of this view were present more than twenty years ago. As Scott McNealy, the CEO of Sun Microsystems, proclaimed to a reporter in 1995, "You have zero privacy anyway. Get over it." McNealy has long been trailed by the remark, held up as either a truth-telling seer or a cartoonish villain of the tech world. What fewer people realize is that McNealy actually held a more nuanced view. Consumer privacy, he believed, was overrated and might as well be dead. In 2015, he said, "It doesn't really bother me that Google and AT&T have information about me, because I can always switch to another provider. If Uber starts screwing around with my data, I'll use Lyft." But privacy as a value is absolutely central to democratic government. If the state violates your privacy, he noted, you can't easily switch governments. "It scares me to death when the NSA or the IRS know things about my personal life and how I vote."

None other than the founder of the World Wide Web, Tim Berners-Lee, echoes McNealy's sentiment. But he also holds companies partly to blame for their willing cooperation in building surveillance technologies for the government. He expresses regret about how far the Web has strayed from his initial vision. After the revelations of Edward Snowden, in an Open Letter to the Internet on the twenty-eighth anniversary of the launch of the web in 2017, Berners-Lee wrote:

> Through collaboration with—or coercion of—companies, governments are also increasingly watching our every move online, and passing extreme laws that trample on our rights to privacy. In repressive regimes, it's easy to see the harm that can be caused. . . . But even in countries where we believe governments have citizens' best interests at heart, watching everyone, all the time is simply going too far. It creates a chilling effect on free speech and stops the web from being used as a space to explore important topics, like sensitive health issues, sexuality or religion.

If the digital age has delivered to us the technologies that constitute a digital panopticon, we have good reason to worry that we have sacrificed the value of privacy entirely. Is there any way out?

FROM THE PANOPTICON TO A
DIGITAL BLACKOUT

In 2009, Stanford graduate Brian Acton and his friend Jan Koum developed WhatsApp, now the most popular messaging app in the world. One of the central tenets of WhatsApp is the preservation of users' privacy through the use of encryption of all messages sent via the app. At its core, encryption is the process of making a message unreadable by anyone other than the intended recipient. Its use dates back to the time of the Roman emperor Julius Caesar, who used a simple form of encryption in his private communications. Since that time, cryptographers—those who study encryption and other means of keeping information secure—have made impressive mathematical and technological advances in maintaining the privacy of communications. In fact, for a time in the 1990s, the US government classified particularly strong forms of encryption as munitions—essentially, certain forms of mathematics were considered to be weapons—and prohibited them from being exported to foreign countries.

Traditionally, the process of encryption involves the person sending the message using an encryption "key" to transform a message (called the *cleartext*) into a form that is unreadable (called the *ciphertext*) to anyone who does not have a decryption key. Only the holder of the decryption key can use it to transform the ciphertext back into readable cleartext. In Caesar's time, the eponymously named "Caesar cipher" was used to encrypt messages by simply shifting each letter in the message backward three positions in the alphabet (wrapping around from *a* to *z*). Conversely, the decryption process involved shifting each letter in the ciphertext forward three letters to regain the cleartext. Today, elementary school children might use something like this as a means of creating "secret" messages to pass to their friends. In the world of digital communication, technologists have created much more powerful and secure techniques for encryption based on complex mathematics. So important is this technology to the field of computing that in recent years numerous cryptographers have received the A. M. Turing Award for their advancement of encryption technology.

WhatsApp transmits messages using a system known as end-to-end encryption: the content of a message is encrypted on the sender's device and decrypted only when it reaches the receiver's device. As a result, no one except the sender and the receiver can read the message—not WhatsApp, not the internet service provider, not anyone who might be trying to tap the communications network—as the message travels over the internet and through WhatsApp's servers while encrypted. The WhatsApp organization doesn't even know any of the decryption keys needed to unlock messages, as those are generated and stored only on users' devices. So if law enforcement comes knocking to ask WhatsApp to unlock messages flowing through its system, it's prevented from obtaining access to them—due not to a policy decision but rather to a technological barrier.

In 2014, Facebook bought WhatsApp for a then-unheard-of price of $19 billion. The acquisition was a way for Facebook to cement itself as the dominant online messaging platform, bringing WhatsApp's 450 million active monthly users at the time, many of them outside the United States, under the Facebook umbrella. Within a few years, both Acton and Koum would leave Facebook citing concerns over user privacy and the way Facebook was monetizing WhatsApp's user base. In turn, in 2019, Mark Zuckerberg, still grappling with the privacy concerns heightened as a result of the Cambridge Analytica scandal, announced an intention to provide end-to-end encryption on all Facebook messaging services, including Facebook Messenger and Instagram Direct.

If you care about privacy, this all sounds appealing unless you are the head of the FBI, trying to track down terrorist sympathizers who are plotting an attack in a major US city or a human rights campaigner in India who has discovered that political gangs are using encrypted communication technologies to organize anti-Muslim violence in advance of an election. Just to underscore the enormity of this challenge: After the siege of the US Capitol and the deplatforming of President Trump, downloads of end-to-end encrypted applications exploded. As the plotters of the insurrection decamped for smaller but completely private messaging platforms, the task of tracking and disrupting the activities of domestic terrorists became far more difficult. How should we weigh the value of privacy against other important benefits? If technologists brought us the tools of encryption to secure communications, perhaps they can provide other technologies that can effectively balance our interest in privacy with our needs for security and safety.

TECHNOLOGY ALONE WON'T SAVE US

Between the poles of Bentham's fully transparent world and WhatsApp's fully opaque world, we might imagine technological solutions that could help us achieve some level of privacy while still making our data available for use in situations that might be beneficial to both individuals and society. For example, in biomedical research, one of the biggest roadblocks is the lack of access to quality health care and medical data. Under laws such as the 1996 Health Insurance Portability and Accountability Act (HIPAA), significant safeguards protect the personally identifiable health care information of patients. Understandably, we might not want information about our current health or medical history to be widely available, including a chronic health condition, any medicines we might take, or the composition of our genetic sequence. At the same time, researchers could use that information, especially when pooled and analyzed, to better determine the efficacy of medical treatments on patient outcomes, the impacts of lifestyle choices on health, or even the design of personalized medications. There would be enormous potential benefits to society—maybe even to you personally—if your health care information were more widely available.

So how can we navigate between these competing aims? One often-touted approach is *anonymization*. The basic idea is that any personally identifying information—for example, your name, Social Security number, street address, and so on—could be removed from any data about you prior to those data being shared with others. As the theory goes, the data that's released would be just about anonymous individuals, nothing that would permit someone to be identified on the basis of the data.

Researchers could still use the data, for example, to correlate various medical treatments with their impacts on patient health, such as determining the efficacy of existing or new drugs on severe cases of COVID-19. They just wouldn't know who the data were originally drawn from.

Though the idea is appealing in the abstract, anonymization is much harder to put into effective practice. That was made clear by Latanya Sweeney, now a professor at Harvard, while she was still a graduate student at MIT. Paul Ohm, a law professor at Georgetown University, recounted the case: "In Massachusetts, a government agency called the Group Insurance Commission (GIC) purchased health insurance for state employees. At some point in the mid-1990s, GIC decided to release [135,000] records summarizing every state employee's hospital visits at no cost to any researcher who requested them. By removing fields containing name, address, social security number, and other 'explicit identifiers,' GIC assumed it had protected patient privacy." Indeed, the governor of Massachusetts at the time, William Weld, went on the record saying that such anonymization would help ensure privacy.

Sweeney saw it differently. She knew that Weld lived in Cambridge, Massachusetts, the same city in which MIT is located. She purchased a list of all registered voters in Cambridge, a document available to the public, for less than $20. Such lists include not only the names and addresses of voters but also their birth dates and genders, attributes that could be matched to records in the GIC data. Indeed, it turned out that there were only six people in Cambridge who had the same birth date as Weld. Three of those six people were men. And exactly one of them lived in the same zip code as Weld. Sweeney now had a way to uniquely identify Weld's medical records in the GIC data by finding those—supposedly anonymized—records that matched his birth date, gender, and zip code. To make a point, she mailed a copy of Weld's personal medical records to his office.

Lest we think that this process of re-identifying (or de-anonymizing) data applies only in rare situations, Sweeney showed in other work that 87 percent of all Americans can be uniquely identified by just three attributes: zip code, birth date, and gender. In fact, the Data Privacy Lab that she founded at Harvard University provides a public website where individuals can check for themselves if they are uniquely identified by those three attributes. Anonymization is a fraught means of trying to protect personal privacy.

A more recent and promising technology known as *differential privacy* aims to provide greater theoretical guarantees on individual privacy, while also making data available for analysis. Originally proposed by Harvard professor Cynthia Dwork in 2006, the premise of this technology is that if two data sets differ only by whether they include the data for a particular individual or not, then the results of asking questions—in the form of computing statistics—using one data set should be almost indistinguishable from the results obtained from the other data set. In other words, whether a specific individual's record is included in the data or not, the statistical results based on that data should be virtually indistinguishable to whoever is asking for those statistics. Note that differential privacy does not allow the underlying data to be directly accessed. Rather, the person trying to use the data (say, a medical researcher) can only ask for certain statistics to be computed from a data set that is otherwise inaccessible.

One way in which differential privacy may be achieved is by injecting small errors—called "noise"—into the statistics that are computed from a data set. To make this more concrete, consider a classic technique in survey design called *randomized response*, which can be viewed as a particular instance of differential privacy. Say we want to ask individuals a sensitive question, such as whether they have ever cheated on their taxes. Naturally, most people won't be inclined to answer this question truthfully for fear of repercussions. So we introduce some "noise" into the way people answer the question to give them plausible deniability—that is, they can potentially claim that their positive (or negative) answer is due to this random noise and not the truth. Here's how it works: After we ask the question but before we get an answer, we ask the person to flip a coin. If the coin lands heads, the person should answer truthfully. If the coin lands tails, he or she should flip the coin again and simply answer "Yes" if it lands heads or "No" if it lands tails. We can't see the result of any flips or know how many times the coin was flipped. So we've given the person a way to answer "Yes" without actually yielding the truth about whether he or she has cheated on their taxes. Moreover, if we tell people that their name will never be associated with their response—we are only counting the number of "yes" and "no" answers we receive—they might be inclined to follow this setup with some reliability.

If we are able to collect this information for a large number of people, we can apply simple statistical techniques to approximate what percentage of people *actually* cheat on their taxes given the total number of "yes" and "no" answers we received, even though there is random noise in the data.

Modern differential privacy often uses more sophisticated techniques than just randomized response, but the underlying principle is the same. It aims to find a balance between preventing data from being individually identifiable and making aggregate information available for a variety of analyses or applications. Instead of asking if people are cheating on their taxes, we could instead ask health care providers about the prevalence of different medical treatments and the outcomes achieved for individuals. We could now carry out medical analytics with less potential for the privacy of health records to be compromised. As patients, we certainly might not want to reveal our personal health records, but we could agree to have the information in them aggregated with other data using differential privacy to enable medical researchers to discover more effective treatments for disease.

In the tech world, Apple and Google have deployed this technology in their products. For example, information about user activity on an iPhone (e.g., words typed, websites visited) can be transmitted back to Apple with a certain amount of noise and no personal identifiers. Apple can then use those data to improve typing suggestions or determine which websites are likely to cause a browser crash without knowing who typed those words or visited those websites. Similarly, Google can determine the likelihood that various ads have been clicked on without necessarily identifying who made those clicks.

Differential privacy seems like a magnificent technical solution to the thorny problem of protecting the data privacy of individuals while still permitting the benefits of innovation via data collection and analysis. But differential privacy is not without its shortcomings. One issue stems from asking a system for a piece of information repeatedly in order to analyze the trend in the answers. Recall the tax-cheating example: If we ask the person using the randomized response scheme only once, we might not be able to glean a great deal of information. But if we ask that person numerous times, we can see a long-term trend toward more "yes" or "no" answers. That will give us greater confidence as to what the true answer is for that individual. In a similar vein, repeating questions to a system employing differential privacy might reveal more information than we expect and belie the notion of how obfuscated or private our personal data actually is. As Sweeney's work on re-identification showed, combining data from multiple sources may enable us to further extract personal information, even when techniques such as differential privacy are employed.

Regrettably, technical solutions alone, though providing some measure of privacy, aren't able to solve our privacy problems in all situations. Moreover, the existence of technologies by themselves doesn't guarantee that they will be used in the best manner for users, if at all. We need to examine the role of technology in a broader context of personal preferences and government regulation.

WE CAN'T COUNT ON THE MARKET, EITHER

Perhaps the best way to manage the trade-off between privacy and safety or between privacy and innovation is by creating a marketplace of different providers. Because some people value privacy more highly than others, we should strive for a competitive market in which private companies develop products and options that offer different levels of privacy protection. This idea fits naturally with one way of thinking about the value of privacy in the first place: the right of any individual to exercise control over information about oneself. It is also consistent with current laws espousing the principles of Notice and Choice that, at least in theory, enable people to make individual decisions on a case-by-case basis about whether to allow for the collection of their data. Of course, the challenging thing about Notice and Choice is that if an individual grants permission, almost any collection, use, or disclosure of personal data is acceptable.

If the marketplace is to balance privacy with other values wisely, three things must be true. The market must be able to deliver a diversity of comparable products with different privacy settings. Individuals must be able to make informed, rational choices about the costs and benefits of different privacy policies. And we must be able to achieve both our personal and societal goals via a set of decentralized decisions we each make about how much privacy we want to protect or enjoy. This is where an idea that sounds so compelling in principle begins to run into trouble.

Let's start with the idea of a competitive marketplace. The best current example of a big tech company deeply committed to privacy is Apple. Indeed, the company's CEO, Tim Cook, has made the battle for privacy a hallmark of his leadership, positioning Apple in opposition to other corporate behemoths in the technology industry. Apple's approach to privacy—"Privacy by design"—privileges the protection of personal data at every turn. The company collects only the personal data required to deliver the services people need ("data minimization"). Its algorithms work with data on your device, so that other people can't see your information ("on-device intelligence"). It provides information on exactly how your information is used, and customers can opt out at any time ("transparency and control"). Information on each device is encrypted, providing protection ("data security"). And individual identifiers are obscured when data must move from your device to Apple servers ("identity protection").

Apple's commitment to privacy runs so deep that the company has even refused requests from federal authorities to cooperate in counterterrorism investigations. The fight over privacy rights broke into public view in 2016, following the San Bernardino, California, shooting that left fourteen people dead. A US court ordered Apple to develop specialized software that would enable the FBI to unlock an iPhone used by the suspected shooter, a presumed ISIS sympathizer. The iPhone was locked with a four-digit passcode that the FBI had been unable to crack, and law enforcement suspected that the phone might contain information about those who had enabled the attack as well as other members of the ISIS network inside the United States. The FBI wanted Apple to create a "back door" that would enable authorities to access the information contained on locked devices—with the appropriate legal authorization. It's almost as though the company was being revisited by the ghosts of the Clipper Chip debate.

In a move that was later called a "bet-the-company kind of decision," Cook refused to cooperate. Though Apple was willing to work with the FBI to try to unlock the phone, it was not willing to create a formal back door, a piece of software that would enable the government to unlock any iPhone in its physical possession. Apple's view was that there would be no way to limit the use of a back door to legitimate, court-authorized circumstances. And the creation of a back door would invite hackers from anywhere in the world to try to find an entry, making all iPhones less secure. Although the FBI ultimately succeeded in accessing the data on the phone without Apple's assistance, the fight between the company and law enforcement demonstrated just how far Cook was willing to go to protect privacy.

Of course, this overarching commitment to privacy over national security is possible only because Apple's business model depends on selling a high-cost, high-volume set of products (phones, computers, tablets) rather than monetizing people's personal information, as Google, Facebook, and others do.

There are other examples of companies that have put a premium on privacy, although relatively few. DuckDuckGo is a US-based search engine that is wildly popular with our undergraduates but largely unknown to most internet users. Its founder, Gabriel Weinberg, described its ambition in relatively plain terms: "We're not trying to topple Google. Our goal is for consumers who want to choose a private option, they should be able to do so easily."

Weinberg's search algorithm amalgamates more than four hundred different sources. Its search results draw heavily on Microsoft Bing, along with Wikipedia, Apple Maps, TripAdvisor, and DuckDuckGo's own web crawler. But the company's unique selling point is its commitment to privacy. DuckDuckGo tracks neither user behavior nor personally identifying information.

Since 2010, DuckDuckGo's growth has been steady but moderate, spurred by growing concerns about privacy. Although its share of the search market remains tiny, the company is profitable, drawing revenue from untracked advertising and affiliate programs. However, the scale of its earnings demonstrates the cost of protecting privacy. An advertising model that doesn't mine the rich behavioral information quietly gathered from users just isn't as lucrative. DuckDuckGo reportedly passed $100 million in revenue for the first time in 2020, while Google earned more than $41 billion in the first quarter of 2020 alone.

The difficulties that DuckDuckGo has faced in trying to grow its share of the search market reveal one of the important reasons—market power—why the market alone won't deliver the privacy-protecting technologies many people desire. In 2018, the European Commission concluded that Google had acted illegally in using its Android platform to drive all traffic to Google Search. The ways Google had done this were not subtle. For example, device manufacturers were required to preinstall Google Search and Chrome apps if they wanted to license the Google Play app store, and Google paid mobile network operators and phone manufacturers directly to preinstall Google Search. The European Union determined that that was an abuse of Google's market position, cementing its dominance in the search industry. Following the European Union's antitrust action, Google agreed to give users in the European market a choice about which search engine they use on Android devices. This will make a big difference in DuckDuckGo's traffic, but it underscores the difficulty of relying on market competition to deliver privacy options when power is concentrated in a tiny number of major firms.

A PRIVACY PARADOX

The second problem is that individuals struggle to meet the challenge of navigating privacy in a digital age. If Notice and Choice is to work as a strategy for balancing privacy and other competing values, people need to have informed preferences about the amount of privacy they want and be able to act on those preferences. Mounting evidence suggests that they don't and can't.

As it turns out, it is really challenging for anyone to develop a thoughtful, reasoned approach to protecting their privacy. Moreover, the potential harms from giving up privacy may be intangible or even largely invisible to most of us. Though the theft of your identity may lead to a long, unpleasant call with a credit agency, the use of your personal data for behavioral ad targeting or algorithmic product recommendations imposes no immediate or visible cost on you. Sharing your daily activities or life history on Facebook, Instagram, Snapchat, or TikTok yields all sorts of good feelings in the near term as you connect with family and friends. But the potential privacy harms to you are difficult to understand and thus rarely weigh heavily on the minds of those who post freely on social media.

Even if you could perfectly forecast the consequences of your privacy decisions, the evidence suggests that people still struggle to formulate stable preferences and act on them. Our students provide a powerful example of this dynamic at work. Every year, when we ask a room full of three hundred students how many of them care about protecting the privacy of their personal data, nearly all of their hands go up. But when we follow up by asking them how many have viewed or made changes to the privacy settings on any of the search or social media applications they use, few hands remain raised. That follows a broader trend; as one study of Facebook privacy settings noted, "36% of content remains shared with the default privacy settings. We also find that, overall, privacy settings match users' expectations only 37% of the time, and when incorrect, almost always expose content to more users than expected."

A colleague at Stanford, Susan Athey, conducted a more systematic exploration of the value of privacy with students at MIT. She surveyed them about their privacy preferences in the context of a campus-wide effort to encourage undergraduates to experiment with the use of Bitcoin. The university offered students a variety of online "wallets" to manage their Bitcoin, each with a different level of privacy protection. The striking fact was that students made choices based on the order in which the wallet options appeared, regardless of their privacy preferences. Athey then went one step further. She wanted to see whether a small incentive could influence the students' decisions about privacy. She and her colleagues offered one group of students a free pizza if they would disclose the personal email addresses of three friends. In a result that is perhaps unsurprising to any parents of college students, they picked pizza over privacy no matter how sensitive they were to the importance of privacy in general.

Social scientists call this the "privacy paradox." There is an enormous gap between what individuals say about the importance of privacy and their actual behavior. It's not just a problem of aligning what you want with how you act. In light of so many intangible or invisible consequences, it is genuinely difficult to determine how much privacy we actually want.

What we see in the classroom, and what Athey saw on campus, isn't really news. Classic work by Alan Westin, formerly a professor of public law and government at Columbia University, classified individuals into three broad privacy segments on the basis of their answers to survey questions: privacy fundamentalists, privacy pragmatists, and privacy unconcerned. When asked directly, most people count themselves in the first segment, indicating a strong desire to control information about themselves. But their behavior turns out to be very different. For example, in one study participants were offered a variety of discounted products by a human-like computerized agent. The agent asked questions of increasing sensitivity as different products were offered, yet few people refused to answer the questions even if they had indicated a strong prior commitment to protecting their own privacy. Anyone who has downloaded a shopping app, joined a frequent-buyer program, or acquired a branded credit card knows that we willingly, even happily, give up privacy about our shopping in return for a few coupons at the store or some cash back at the end of the month.

In a world in which it is challenging to figure out just how much you care about privacy, the natural tendency is to let context or other influences shape your choices. But the fact that privacy preferences are so context dependent and malleable is yet another reason to be concerned about the limitations of Notice and Choice.

Take the American public's views on the protection of personal data in the aftermath of 9/11. In the years immediately following the attack, citizens were concerned that the government was prioritizing civil liberties too much and not going far enough to protect the country from another terrorist attack. Edward Snowden's revelations about mass surveillance flipped people's views almost overnight. Within six months, the share of Americans who supported the government's collection of telephone and internet data as part of antiterrorism efforts dropped precipitously.

We've seen similar dynamics at play during the COVID-19 pandemic. As the disease was spreading rapidly in May 2020, though the American public was broadly (more than 70 percent) concerned about the security of its personal information and felt that the situation was getting progressively worse, a public health pandemic was enough to generate majority support for the government to use cell phone apps to track the location of infected individuals as part of a digital contact-tracing effort.

People's decisions about how much to disclose about themselves are also influenced by what others say and do. In some cases, this might be relatively harmless, such as when friends provide a current photo of their family on Facebook and encourage others to do the same. The challenge is that the business model for many tech companies depends on the collection of personal information to drive higher profits. These companies often exploit people's amorphous privacy preferences in predictable and systematic ways. Default settings are a good example. Even if the default is set to maximize the disclosure of personal information, many individuals are unlikely to change those settings, regardless of their privacy preferences. Sometimes they don't know how to or even realize they can.

There are even more pernicious strategies that exploit human vulnerabilities, such as the unreadable language in privacy policies, malicious design features that frustrate the ability of users to change settings, and the manipulation of our perceptions of control that actually induce people to disclose even more information.

The current practice of Notice and Choice does not work in a world in which people are unaware of what they are sharing and how it can be used. Even when they have that information, they still struggle to make reasonable judgments about how much they want to protect their privacy. Moreover, when companies have strong incentives to manipulate how much personal information people disclose, it's even less likely that individuals will be able to navigate this terrain on their own.

PROTECTING PRIVACY FOR THE BENEFIT OF SOCIETY

The third challenge is that we may sacrifice other things we care about if we leave it to individuals to make their own decisions about privacy. The benefits and costs of privacy protection are better evaluated "cumulatively and holistically," according to George Washington University professor of law Daniel Solove, given the social consequences of the individual decisions we make.

This is particularly true when it comes to data collection that serves the public interest. Public institutions often need to be able to access personal data and be able to learn from that data, even if access might make particular individuals uncomfortable. A veritable alphabet soup of abbreviations—the Fair Credit Reporting Act (FCRA), Health Insurance Portability and Accountability Act (HIPAA), Family Educational Rights and Privacy Act (FERPA)—speaks to the seriousness with which policy makers have sought to balance privacy protections with the government's need to access information on people's credit, health status, and education records.

The COVID-19 pandemic has brought these issues into sharp relief.

Effective strategies of pandemic surveillance have the potential to bene-fit all of us as we confront a once-in-a-century public health crisis. But whereas the Chinese government is comfortable with mandating access to people's personal information, in a democracy, the growth of digital contact tracing depends on balancing the public health benefits against people's concerns about privacy.

The efforts by Apple and Google to develop a COVID-19-tracing technology demonstrate this careful balancing act at work. If you opt into the system, your phone will let you know when you have been ex-posed to someone who has tested positive for the virus—as long as that person has also downloaded the app and truthfully updated his or her health status. As an alternative to traditional modes of contact tracing—the slow, patient work of manually listing people who have tested positive and visiting them in their homes—this is a revolutionary approach.

Keenly aware of the fact that many people are distrustful of tech companies, Apple and Google have built privacy features into their contact-tracing apps to try to ensure compliance. For example, the tech-nology doesn't collect any location data. Instead, it relies on proximity data gathered through Bluetooth without using any other information about you from your phone. The system can tell you that you've crossed paths with someone who has been exposed to the virus but protects that information on your phone and does not share it with Google, Apple, or any government agencies. This commitment to privacy gives public health officials and citizens a good shot at enrolling large numbers of people in digital contact tracing, but the success of tracing ultimately depends on people's willingness to enroll and update their health status.

Oftentimes the social benefits of disclosure are too important to risk relying on people's willingness to share their private information. The na-tional security space is where this dynamic is most obvious. We can't de-pend on the decisions that terrorists voluntarily make about the disclosure of their personal information to keep the country safe. Instead, democra-cies have created rigorous legal and judicial processes that determine who can violate people's privacy and with what legitimate authorization when there is a credible threat to safety and security. The same goes for efforts to tackle child pornography, human trafficking, cybercrime, intellectual property violations, and other social ills that would be enabled—and even amplified—in a world that protected privacy above all else.

There are lots of reasons to be skeptical that the market will help us achieve a good balance of privacy and other social goals. Though the idea is appealing, an approach that depends on informed individuals making good judgments about their privacy preferences and choosing from a diversity of products with different privacy protections is fanciful. Taken together, the nature of the market, the limits of human decision-making, and the obvious social benefits (in some cases) of accessing private information make clear that our current approach to regulating privacy is incomplete and even untenable. We need a new direction.

FOUR LETTERS THAT ARE KEY TO YOUR PRIVACY

Depending on who you talk to, the letters GDPR either signal intrusive government regulation gone terribly awry or exemplify the twenty-first-century version of the fight for civil rights in a digital age. The General Data Protection Regulation (GDPR) is the European Union's attempt to bring order to the battle over personal data.

Europe has always been more interested in data protection than the United States has. Perhaps this is a legacy of its authoritarian past, its citizens haunted by the memory of the ways in which Nazi Germany and the Soviet Union used mechanisms of surveillance and information control to silence opposition, repress citizens, and carry out unspeakable crimes. In the 1930s, German census workers went from door to door, gathering data on residents' nationality, native language, religion, and profession. Using machines manufactured by IBM's German subsidiary, the government collated the information and used it to organize and implement its Holocaust against Jews and other marginalized groups.

State surveillance continued in East Germany and in other countries behind the iron curtain after the war. In tactics made famous by the Oscar-winning movie *The Lives of Others*, the secret police (the Stasi) bugged bedrooms and bathrooms, read mail, and searched people's homes at a moment's notice, destroying any distinction between private and public life. Given this history, it's no surprise that in 1983, Germany's Federal Constitutional Court declared "self-determination over personal data" to be a fundamental right—one that was extended to East Germany after the Berlin Wall fell in 1989.

But the recent push to update data privacy regulations has more contemporary roots. When Edward Snowden revealed classified details on the work of the NSA, Europe was outraged. There were countless embarrassing allegations for the United States, including evidence of spying on 122 high-ranking world leaders, including German chancellor Angela Merkel via her mobile phone and some 70 million French citizens via their phone calls and emails. But most disturbing for many Europeans was the discovery that private technology companies, including Apple, Yahoo!, and Google, had been forced by governments to hand over user data to US and UK intelligence agencies.

In the wake of the revelations, Jan Philipp Albrecht, a thirty-year old German member of the European Parliament, announced himself Snowden's "comrade-in-arms." Although his tactics differed, Albrecht saw himself as engaged in the same battle: a war against US government surveillance. Building on a prior policy directive, the little-known Green Party member authored a new set of rules to protect EU citizens from wholesale snooping and to impose enormous fines on private companies that collect and hand over data without a user's permission. Although stringent rules had been the subject of fierce debate before the Snowden leaks, it was no longer politically wise for any lawmaker to oppose sweeping new safeguards for online data in Europe. Albrecht championed data protection in Europe as "part of our self-determination and dignity," and the new rules passed a preliminary vote after only two hours of discussion, beginning a long process that culminated in the adoption of the GDPR in 2016.

GDPR is substantially different from the US Notice and Choice approach. Instead of relying on individual users to make informed judgments about the terms and conditions they are willing to accept, the GDPR adds distinct legal criteria under which personal data can be collected and creates a set of affirmative consumer rights to which every company serving EU residents is bound. For example, personal data may be lawfully collected on the basis of consent or to perform a service, but also for other reasons including "where it is necessary to protect an interest which is essential for the life of the data subject or another individual" and for tasks carried out in the public interest. This creates the ability for the government to collect and gain access to personal data, but only when necessary.

More important, EU residents are entitled to a new set of digital rights. Companies must inform users that their data are being collected, the purposes for which the data are being collected, and the entities the data will be shared with. Users have a right to obtain a copy of their data, to correct their data, to erase their data, and to remove their data from one company and give it to another. GDPR also empowers residents to constrain how companies use their data. Users can stop companies from processing their data and can object to the use of automated tools that make decisions relevant to their employment, economic situation, health, and welfare.

GDPR is a modern bill of rights for the digital age. And it can't be dismissed as simply a set of legalistic aspirations. The authors of GDPR were as serious about enforcement as they were about creating new rights. In fact, the penalties associated with violating GDPR are potentially enormous. A country's data protection authority can assess penalties of up to 4 percent of a company's global annual revenues—a huge fine for the world's largest tech companies.

This bold move by the Europeans sent shock waves across the Atlantic. With its focus on all companies that conduct business with EU residents, GDPR sets standards that apply to all major tech companies regardless of where they are headquartered. User complaints are already pouring in to data protection authorities (more than 95,000 complaints filed with the European Data Protection Authorities as of 2019), and major suits have been filed against Facebook, Google, Instagram, and WhatsApp in Ireland. US technology companies have had to take notice. Days after GDPR went into force in 2018, Mark Zuckerberg commented on Facebook's plan to comply, "We're still nailing down details on this, but it should directionally be, in spirit, the whole thing." After coming under significant criticism, he cleaned up his comments a few days later, saying, "We're going to make all the same controls and settings available everywhere, not just in Europe." And so was born the European Union's current status as the world's de facto tech regulator, especially on issues of privacy. But it might not last very long.

BEYOND GDPR

Facebook was waking up to a new reality, not only in Europe but also in its own backyard—because just as GDPR was coming into force, California was moving forward with its own revolution in privacy law, the California Consumer Privacy Act (CCPA).

The story begins around 2016 with a real estate billionaire Alastair Mactaggart, who asked a friend, a software engineer at Google, whether he should be worried about the personal data Google had about him. He hoped for the comforting answer given by an airline pilot when asked about the potential for plane crashes. Instead, his friend raised his level of anxiety substantially: "If people really knew what we had on them," the Google engineer said, "they would flip out."

Mactaggart, who had grown wealthy by building condominium developments for Silicon Valley's young, successful techies, became curious about the role of data, the fuel driving the growth of Google, Facebook, and many others. He quickly discovered that the United States had no single, comprehensive law governing personal data. He grew more disturbed as he learned that the patchwork of regulations that did exist had largely been designed by the very companies that were benefiting most from the monetization of people's personal information. His journey of self-discovery revealed something that is now widely understood: if you're not paying for the product, you *are* the product. The real customers of many of the big tech companies are the advertisers, and what they're paying for is the ability to target their messages ever more precisely to consumers, based on their interests, preferences, and desires.

Mactaggart was incensed. He learned that the Obama administration had proposed a "consumer-privacy bill of rights" in 2012 that had fallen victim to the Snowden revelations as the government had lost its momentum and moral authority to drive reforms on data protection. Even as government officials sought to resurrect legislation in Obama's second term, the tech industry mounted a furious lobbying campaign. Big tech convinced lawmakers to gut the bill, forcing a retreat from the idea of consumer privacy as an inherent right. By the end of the process, the consumer advocacy groups hated the final product, as did the tech industry—which, despite Zuckerberg's statement, had never wanted the legislation in the first place.

Fast-forward to 2018. Mactaggart envisioned an end run around the tech companies via California's peculiar effort at direct democracy, the ballot initiative process. This is the same ballot process that Uber and Lyft used to pass Proposition 22 in 2020, the provision that exempted app-based transportation and delivery companies from providing full benefits to their employees. Almost anything can be taken directly to voters, bypassing the legislature, if petitioners can gather enough signatures. Mactaggart and his allies drafted a ballot initiative that was narrower in scope than GDPR and the Obama-era Consumer Privacy Bill of Rights. They went straight for the most egregious aspect of Silicon Valley's hidden use of personal data: the collection and sharing of personal data with third parties. The ballot initiative empowered California consumers to learn what personal information companies were collecting and how they were using it and gave them the option of preventing the selling and sharing of their data.

Mactaggart's ballot initiative garnered 629,000 signatures, double what was necessary. Enthusiasm for the effort mounted as Facebook confronted a scandal around Cambridge Analytica's misuse of the personal data of 87 million people in 2018. By that point, legislators had taken note and the companies were far more open to compromise. Mactaggart agreed to hold off on his push to put the initiative onto the ballot, hoping that the state legislature could develop a comprehensive privacy bill.

In June 2018, Governor Jerry Brown signed CCPA into law. California was following in the footsteps of the European Union in committing itself to the protection of privacy. And although CCPA is not as ambitious as GDPR in many ways, some core elements remain the same: a requirement that individuals must consent to the way in which their data are used; the ability to prevent companies from collecting and selling or transferring their data to others; and the right to demand that companies delete their personal information. CCPA contains solid enforcement mechanisms, including through what's called a "private right of action"— the ability of individuals who have been wronged by a company to sue in court—though lobbying by the big tech companies watered down this provision in the final draft. Also like GDPR, CCPA applies to any companies that do business in California, not only companies based there, expanding its territorial reach. Now energized about a right to privacy, Californians overwhelmingly passed a new ballot measure in 2020, further strengthening CCPA, including the establishment of the California Privacy Protection Agency to enforce the law.

● ■ ● ■ ●

It is too soon to draw firm conclusions about GDPR and CCPA. As the new frameworks face legal objections and implementation challenges, they nevertheless portend a wholesale transformation in how people think about privacy in the digital age. The idea that we can rely on individuals to protect their own privacy when the world's largest companies have every incentive to gather, analyze, and sell as much personal information as they can possibly collect seems old-fashioned. Notice and Choice, on its own, is becoming a relic of the past. The question is where the United States and other democracies will go next. Is comprehensive privacy legislation a real possibility? If so, what must it accomplish?

At a time when the tech companies are as unpopular as they have ever been, perhaps it's not surprising that the public overwhelmingly supports comprehensive federal privacy legislation. And politicians have taken note. The House of Representatives and Senate are swimming in draft privacy legislation. But the proposals take significantly different approaches. Some bills seek to nationalize GDPR- and CCPA-like requirements in the United States and create more powerful enforcement mechanisms, while others seek to preempt state laws with a national data privacy standard that is far less constraining (and thus more responsive to corporate concerns).

Though the detailed negotiations will be left to the politicians, the outcomes of the political process should reflect what we citizens want—not just what corporations and their lobbyists want—in a twenty-first-century balance of privacy against other values. And to us, it's clear what any new privacy legislation will need to achieve.

First, we need to move aggressively in the direction of treating privacy as a right, not just a preference that individuals have in some contexts but not in others. That was the instinct of policy makers in the Obama administration as they imagined a consumer privacy bill of rights, and it is reflected in the significant legislative achievements in Europe and California. By defining this right in detail—and being concrete about what it entails and how it must be protected—legislation would bring an end to the Wild West mentality that has governed the use of people's personal information in the tech sector. It would shift power from companies back to citizens, and users of tech products would regain significant control over the data they provide and how they are used. This doesn't mean abandoning our commitment to giving consumers choice. It does mean resetting the baseline expectation about what's okay and what is not and creating an expectation that companies need to do more than seek uninformed and meaningless consent from users via terms of service that most people don't read or understand.

The second element is a commitment to more meaningful forms of consent. Some of this can be accomplished via rules. Companies should no longer be able to take advantage of consumers' ignorance to harvest as much data as possible. As under GDPR and CCPA, citizens should have a right to know exactly what information companies are collecting and how they will use it. It should no longer be acceptable for platforms to present the details of what and how data will be used in ways that no reasonable individual is patient or well informed enough to understand. Companies should no longer have license to gather up data on just about anything. Instead, principles of data minimization—extracting just what is needed and explaining why—should govern corporate behavior.

But as Apple's approach to privacy or new technologies suggests, the design of tools can also protect privacy rights and make it more likely that individuals will be able to act in line with their preferences about what information they wish to share or keep private. For example, a shift from a default setting of opt out, which the companies prefer, to opt in could have significant implications for the personal data users share. This change would make it easier for privacy-sensitive individuals to achieve their goals. To be fair, though, it would come at the cost of reams of data that companies now use to personalize services, target advertisements, and improve their products. Whether having to opt in will create a better balance is an open question and one that would benefit from active experimentation. But it does point to the value of "nudges" built into products. Nudges are design features that provide an architecture for people's choices without constraining their options or changing their economic incentives.

Another tool that could make an enormous difference is one that would enable users to set their privacy preferences and take them from one platform to another, instead of having to do so on each platform or site. Some call this approach user-centric privacy. In the same way that we attach our preferences to a standardized telephone number or address, we should be able to do the same with a neutral online identifier that carries our privacy preferences. A number of entrepreneurs are racing to figure out how to do this. For example, a former Google executive, Richard Whitt, is designing a "digital trustmediary" that will sit between users and platforms to control what data are shared at any moment. And the internet pioneer Tim Berners-Lee has a new start-up called Inrupt that organizes users' data into "data pods" that grant applications access to their data selectively and in line with the users' preferences.

Third, we need a credible, legitimate government agency that has the power to preserve and protect the privacy rights defined by new legislation. Users should not be responsible for micromanaging their own privacy settings, just as consumers are not responsible for determining whether a car is safe to drive or food will make them sick. Consumers should be able to relax, knowing that someone is looking out for their privacy and setting reasonable standards that protect their basic rights.

Much of the debate on institutional design in the United States relates to the role of the Federal Trade Commission (FTC). An organization with the mandate to counter "unfair and deceptive business practices" has become the default privacy regulator in recent years, policing companies to ensure that they honor their own privacy promises. The FTC began regulating online privacy in the mid-1990s with a set of enforcement actions narrowly tailored around children's privacy—for example, going after KidsCom.com and Yahoo! GeoCities for collecting information from children and sharing it without parental consent. More recently, it has targeted companies that fail to seek user consent for changes to their privacy policies. The FTC's first unfairness ruling, against Gateway Learning Corporation, the creator of the children's series Hooked on Phonics, held the company accountable for retroactively changing a privacy policy to allow the selling of user data to third parties. The FTC also ruled that Facebook had deceived its users when it shared the data of users' Facebook friends with third-party app developers, even when those friends had more restrictive privacy policies in place.

But despite all of this activity, critics of the FTC call it "low-tech, defensive, toothless." One of the major advocacy groups, the Electronic Privacy Information Center (EPIC), argues that the FTC can't do much in a world governed by Notice and Choice, as it is limited to reactive and after-the-fact enforcement actions that ensure companies honor what they say they will do in their unreadable and mostly unread privacy policies. The bottom line is that a new set of rights will be meaningless unless the US creates a purpose-built agency with the power and resources to make rules, monitor implementation, conduct investigations, and sanction entities that violate users' privacy rights. The US needs a "public trustee" whose express mandate is to protect people's personal information, while balancing privacy against our other shared goals. When introducing a legislative proposal for a data privacy agency in February 2020, Senator Kirsten Gillibrand pointed out that "the United States is vastly behind other countries on this. Virtually every other advanced economy has established an independent agency to address data protection challenges." It's time for the United States to follow suit.

● ■ ● ■ ●

We live in a surveillance society. The technological wonders of the digital tools we use every day enable governments and companies to know more about what we want, what we do, and what we think than ever before. Although Americans have had a long history of skepticism of government access to their data, they increasingly have a healthy suspicion of private companies harvesting too much data as well. Tech companies profit from our personal information and turn access to data about each of us into products that make them rich.

Moments of social mobilization bring many of these concerns to the surface. When our rights to free speech and free association are threatened by governments that carry out aerial surveillance of marches or by private companies that use facial recognition tools to help the police track down those they accuse of misdeeds, the public demands greater protection of its privacy rights. But until we define those privacy rights clearly and create an institution that has the power and authority to protect them, we'll be left in the void of Notice and Choice—a place where we feel as if we have some control over access to information about ourselves but where, in reality, companies take advantage of our ignorance, laziness, and cognitive limitations to gather ever more information on us. There is an alternative destination: a place where we don't need a high level of tech proficiency to protect our basic rights.

CHAPTER 6

CAN HUMANS FLOURISH IN A WORLD OF SMART MACHINES?

Humans have been fascinated by the idea of self-driving cars for decades. Yet for many years that reality was frustratingly out of reach. As with so many innovations in computers, it was the US Department of Defense that produced a breakthrough.

In 2002, the Defense Advanced Research Projects Agency (DARPA), announced the DARPA Grand Challenge, a 142-mile car race in the Mojave Desert. The race had a particularly intriguing requirement: the vehicles were to be entirely autonomous—that is, after starting the race, humans could not intervene in their operation. The prize: $1 million and a place in history. On March 13, 2004, amid the high hopes of the fifteen entrants, those gathered at the track watched vehicle after vehicle veer off track and get stuck in the desert sand. Some vehicles barely made it past the starting line. The vehicle that progressed the farthest in the race—a Humvee named Sandstorm engineered by Carnegie Mellon University's race team, which was the early favorite—traveled a little over seven miles (less than 5 percent of the total race distance) before going off course and lodging itself on a berm. The race was considered a failure, and no prize money was awarded.

But DARPA officials refused to give up. They scheduled a second race for the following year and doubled the prize to $2 million. The 2005 DARPA Grand Challenge was a wholly different story. Twenty-three teams qualified for entry on a new 132-mile-long desert course, with five vehicles successfully completing the entire race. The winner was a car named Stanley, fielded by a team from Stanford University, led by computer science professor Sebastian Thrun. Stanley completed the course in 6 hours and 54 minutes, beating the second-place finisher by eleven minutes. Astonishing progress had been made in just eighteen months. The race, which had been seen as a dismal failure the previous year, was now touted as a resounding success.

Reflecting on the race, Thrun crowed, "It's a no-brainer that 50 to 60 years from now, cars will drive themselves." His estimate was far too pessimistic. In 2020, at least thirty different countries were testing autonomous vehicles on their roads. California alone licensed more than fifty companies and more than five hundred autonomous vehicles, logging more than 2 million miles.

One of the primary motivations for the commercial development of autonomous vehicles is their potential for improving road safety. The World Health Organization estimates that there were 1.25 million road traffic deaths globally in 2013. In 2017, 37,133 people were killed in motor vehicle crashes in the United States alone, and more than 90 percent of those crashes involved human error. Of course, in assessing the question of safety, we need to ask how safe is safe enough for driverless cars to become the norm. It's important to remember that a driverless car need not be perfect, it just needs to be better than a human driver. And as the statistics indicate, humans aren't such great drivers—we are error prone, sometimes piloting vehicles while overly tired, distracted by text messaging, or drunk. Numerous other arguments have been made for the benefits of driverless vehicles, including making commuting time more productive (estimated to save hundreds of billions of dollars annually in the United States alone), reducing fuel consumption by having cars "platoon" or flock together by coordinating their movements, increasing the throughput of existing roads (since self-driving cars won't need as much space separating them while they drive), and even reducing the need for parking spaces.

So what could possibly be the downside of deploying driverless cars? Well, first we need to figure out whether they work, and that turns out to be a tough question to answer. Beyond demonstrating that driverless cars can make low-level decisions, such as following the rules of the road and avoiding accidents, we need to understand how they make more complex high-level decisions. For example, when confronted with a choice about whether to swerve into a bicycle lane to protect the car's driver or to harm parents bicycling with their children, what should the autonomous system that pilots the car be programmed to do?

Consider a hypothetical dilemma introduced by the English philosopher Philippa Foot in the late 1960s, the "Trolley Problem," that has now become a real problem for engineers. In the context of autonomous cars, the problem asks whether a vehicle should be programmed to endanger or sacrifice the life of its sole passenger by running off the road in order to avoid potentially hitting five pedestrians crossing the road. As a society, we might prefer that the vehicle pick the option that saves the most lives. Accordingly, most participants in a 2016 survey preferred that such vehicles be programmed to minimize casualties on the road generally. But the same survey showed that when participants were asked to be a passenger—that is, potential purchaser—of an autonomous vehicle, "they would themselves prefer to ride in AVs that protect their passengers at all costs." They reported that they would be less likely to purchase an autonomous vehicle that did not prioritize passenger over pedestrian safety, yet they would like others to buy them. Such conflicting views of societal versus personal preference in purchasing decisions can lead to bad outcomes. Seen from the perspective of a pedestrian or bicyclist, we prefer to prioritize the welfare of all people. Seen from the perspective of a potential driver, we prefer to prioritize passenger safety. Autonomous systems can't be programmed to do both at the same time. It's an indication that a free-market approach here won't get us the results we're hoping for.

Though people are beginning to accept that a world of driverless cars is just around the corner, we must also prepare for the automation of other functions in which the displacement of human judgment just doesn't seem right. These might include the role of a doctor in diagnosing disease and prescribing the right treatment for you, the care with which a teacher evaluates the needs of your child and provides lessons to advance his or her skills, and the authority we grant politicians to make life-or-death decisions about how to protect our national security. These roles demand empathy or human agency connected to the outcomes.

There are amazing possibilities for computers to save lives on the road, in the doctor's office, and on the battlefield. But some of those possibilities come with costs that may not be easily measurable—and thus cannot be optimized for. A future is coming in which we will have less agency over our lives, the benefits and burdens of automation will be unevenly distributed, and we may even find ourselves replaced by the technologies we can now create. Now is the time to contemplate these repercussions.

BEWARE THE BOGEYMAN

Science fiction writers have long conjured up worlds where machines have become more intelligent than humans and humans become subservient to our new robot overlords. In our present age of massive advances in artificial intelligence, there are already domains in which machines can reliably outperform humans. But is this something we should worry about? The answer is complicated, and how it plays out will depend on the kind of tasks we give over to machines and the way we design the interactions between machines and humans.

One conventional way to trace the history of AI is by its success in defeating humans at various tasks. In 1959, Arthur Samuel, who popularized the term *machine learning*, developed a computer program that played checkers and could improve—could seemingly "learn"—by playing against itself. It was not until 1994 that a computer program could defeat the human world checkers champion. Then, in 1997, front-page stories in newspapers around the world announced that IBM's Deep Blue computer had "unseated humanity" by defeating the reigning chess champion, Garry Kasparov. In 2011, IBM's Watson system, built to play the TV quiz show *Jeopardy*, soundly defeated the two former all-time winners, Brad Rutter and Ken Jennings. Final score: Rutter with $21,600, Jennings with $24,000, and Watson with $77,147. In 2017, scientists from Google's DeepMind group used machine learning to build a program that would go on to beat Ke Jie, the number one Go player in the world. Until that time, many game players believed that Go, a game that has far more than a billion billion billion more board configurations than chess, was beyond the scope of world-championship, or even expert-level, play by a computer. The success of AlphaGo left many of those gamers startled, with some players reporting the computer's play as "from another dimension."

These are astonishing technical accomplishments. And they're just the beginning. Rapid advances in artificial intelligence and its deployment in autonomous systems herald an age in which machines can outperform humans at more than just games. On the one hand, the automation of unpleasant, dangerous, alienating, or mind-numbingly repetitive tasks has already proved to be a blessing for many. On the other hand, the automation of other kinds of tasks from which we humans derive meaning, pleasure, and fulfillment may become a curse and herald a transformation in the very meaning of what it is to be human.

Do you enjoy doing laundry? Almost no one does, and this despite the fact that doing laundry today is easier than it ever has been. Before the advent of washing machines, cleaning clothes was a time-consuming process of soaking, soaping, rinsing, and drying each garment. The washing machine is nothing more than an automated system that relieves the need for human labor. And to the great benefit of humanity! Washing machines not only liberated humans from the boring, labor-intensive task of doing laundry but, given the traditional division of labor in households around the world, freed women in particular from drudgery. Does anyone lament the diminished place of human labor in the task of doing laundry? We doubt it.

Washing machines are not especially intelligent, but they are more effective and much more efficient than humans. The same is true of many other machines developed over the course of history, from the combine harvester to the calculator. The automation of tasks previously performed by human labor long pre-dates artificial intelligence. In the late 1800s, more than 50 percent of jobs in the American economy were in agriculture. By 1930, the figure was 20 percent, and in 2000, it was below 2 percent. Such transformations of the labor market can cause painful disruptions and social upheaval, to be sure. But what economists call the technological displacement of labor or, less formally, the loss of jobs in a sector of the economy due to the introduction of machines, is a familiar feature of modern life. And it's not just as a result of AI.

The crux of the debate about AI is not in handwringing about the rise of robot overlords, but in considering the kind of work that machines are replacing and how we are to manage this transition. If the work is more complex than doing laundry and contributes to our well-being, we might lament the loss of human labor.

Is there something special about our age of computer-powered machines and digitally driven automation? Early forms of automation tended to substitute machine for physical labor. But digital automation can replace both physical and cognitive labor and change our experience of work. And unlike physical devices or machines, digital automation is relatively cheap to duplicate. But even if digital automation is more powerful than any previous form of automation—because it can substitute for human thinking and outperform humans at information processing— we need not worry about smart machines rivaling human thinking and reasoning. At least not yet.

There are important differences between human and machine intelligence. Human intelligence involves the twofold capacity to reason about our goals as well as the means to fulfill them. Humans, perhaps uniquely among all living creatures, can reflect upon and revise their most fundamental aims in life. Machines, even the most intelligent, can neither set their own goals nor deliberate about whether their goals are worthy or unworthy. It was humans who decided to program a computer to play checkers or chess. It was humans who decided to use computers to identify faces in images. Technologists are especially good at creating smart machines when the rules of a game are unambiguous, including what it means to win or lose. It's no accident that the catalogue of milestones in artificial intelligence includes a great many games.

What's harder is creating intelligent machines that strive to accomplish goals that are more difficult to define. What, for example, is the goal of an autonomous vehicle? What objective function should the autonomous system, programmed by a technologist, try to optimize? Some might say the goal is the fastest transport of passengers to their destination. Others might say it is the safety of the passengers during the journey. Others still might prefer that the ride be enjoyable, perhaps by driving on scenic rather than speedy routes. Whatever the objective function, the autonomous system has to be capable of navigating an unpredictable and changing environment: different kinds of roads, weather, and lighting conditions; the behavior of other cars and pedestrians; and surprises such as obstacles in the roadway, including children chasing balls into the street. Until machines are capable of defining their own goals, the choices of the problems we want to solve with these technologies—what goals are worthy to pursue—are still ours.

There is an outer frontier of AI that occupies the fantasies of some technologists: the idea of artificial general intelligence (AGI). Whereas today's AI progress is marked by a computer's ability to complete specific narrow tasks ("weak AI"), the aspiration to create AGI ("strong AI") involves developing machines that can set their own goals in addition to accomplishing the goals set by humans.

Although few believe that AGI is on the near horizon, some enthusiasts claim that the exponential growth in computing power and the astonishing advances in AI in just the past decade make AGI a possibility in our lifetimes. Others, including many AI researchers, believe AGI to be unlikely or in any event still many decades away. These debates have generated a cottage industry of utopian or dystopian commentary concerning the creation of superintelligent machines. How can we ensure that the goals of an AGI agent or system will be aligned with the goals of humans? Will AGI put humanity itself at risk or threaten to make humans slaves to superintelligent robots or AGI agents?

However, rather than speculating about AGI, let's focus on what's not science fiction at all: the rapid advances in narrow or weak AI that present us with hugely important challenges to humans and society.

WHAT IS SO SMART ABOUT SMART MACHINES?

How did our machines get so "smart" that they now threaten to replace their creators in some lines of work and cause us to question our own human abilities in other tasks? Understanding how and why we came to stand on the precipice of a transformative AI future takes us back to the post–World War II era. The path of development was not smooth or incremental.

The term *artificial intelligence* was first coined in 1956. Computer science was just becoming a field of its own, and the possible futures that computing enabled seemed endless. Early advances in AI research in the 1950s through 1970s, such as Arthur Samuel's checkers program and an "expert system" named MYCIN that could diagnose blood infections better than human doctors, led to great optimism about what could be accomplished. But great enthusiasm in the lab is often followed by great hype in the business world, and AI came to be trumpeted as a potential business solution for a whole host of problems, many of which were far beyond what the technology of the time could achieve.

Much of the early work in AI tried to model deduction and human decision-making through logical rules, such as "All dogs have four legs." But such rules-based systems are brittle—what happens to your logic when you encounter a dog that is an amputee and only has three legs? A rules-based system might then infer that it isn't a dog. Or worse, it might enter an inconsistent state where it couldn't reason at all because it was given information that wasn't consistent with its existing rules. Attempts to modify the rules in such systems quickly become mired in endless minutiae that try to address exceptions: What if the three-legged dog has a prosthetic leg? What if the dog's prosthetic is actually a wheel? And so on. Such rules-based systems become complicated, slow, and failure prone.

This example of an amputee dog may not look like a significant problem, but if inadequate rules-based systems are used in critical situations, such as controlling a blast furnace in a steel mill, problems in reasoning have the potential for much more significant consequences.

The overhype and underperformance of AI systems led to a decades-long "AI winter" in the 1970s and '80s. Industrial use of AI was significantly curtailed, and academic funding dried up.

Toward the end of that period, more modern AI systems overcame the brittleness of rules-based systems with the development of more flexible and often complicated systems. These included the advent of neural networks, algorithms for machine learning that took inspiration from models of the human brain as a way to deal with uncertainty in the world. Absolute rules were replaced by numbers, often representing probabilities and possibilities, which could be mathematically combined in robust ways that allow systems to handle unexpected situations without a complete breakdown in the ability to reason.

Today, AI systems are much more flexible and powerful in the way they can model and make inferences about the world. They are able to make cars navigate city streets, recognize faces and identify who is pictured, and play board games at levels that beat world champions. But the power of these newfound systems comes with a cost: complexity. These AI systems are now so complicated that the humans who program them have difficulty understanding why some machines make the decisions they do. AI systems can identify patterns in huge pools of data that humans can't discern and can therefore frequently make more accurate predictions. But such systems are often black boxes, unable to explain why they generate particular outputs. And the scientists who build the systems can't always explain the outputs either, making the decisions inscrutable.

Deep learning refers not to the capacity to generate insights but to a spatial metaphor for the architecture of the AI system. The idea behind deep learning is that inputs to a system form sets of simple patterns that are then combined into increasingly more complex patterns using patterns derived from the previous layer. The moniker *deep* comes from the fact that such systems now contain many more layers than they did just a decade ago—a result of greater computational power to model the patterns in each layer and vast increases in data that enable these more complex patterns to be discovered. For example, in a facial recognition system, the input is generally a picture (a "faceprint") or, more specifically, a grid of pixels that compose the image. The first layer in the system might identify lines, or "edges," from the underlying pixels. In the next layer, the lines are combined into simple shapes that are in turn combined in a subsequent layer to represent facial features such as eyes, noses, and mouths. A final layer may then combine these facial features in the right configuration to represent a face. It sounds simple enough, but actually building such a system is an enormous undertaking and can be accomplished only using machine learning techniques. Facebook's DeepFace system for facial recognition uses a "nine-layer deep neural network. This deep network involves more than 120 million parameters." And that was in 2014. Since then, tech giants such as Microsoft, Google, and NVIDIA have built models with tens of *billions* of parameters, and the numbers continue to grow every year. Such models are beginning to approach the complexity of the human brain, which is estimated to have roughly 100 billion neurons. The question of what deep neural networks are really computing far outstrips the human ability to comprehend all the parameters involved.

The power of AI systems has spurred a dizzying array of applications in our lives. Some of them we've used for years, perhaps unknowingly. Methods to filter spam from our email or comb our credit card purchases for possibly fraudulent transactions have been used for a couple of decades. If you get a call from your credit card provider notifying you of suspicious activity on your card, you can most likely thank AI for finding it. But in cases such as spam filtering and credit card fraud, the volume of transactions is far greater than humans can process. The use of these technologies didn't displace any workers because the tasks were not something that humans could keep up with in the first place. And they were hidden in the bowels of data systems, not a place where workers are found.

As AI grows in its pattern recognition capabilities, it is able to power more complicated tasks, oftentimes in more visible ways. Autonomous vehicles are just a start. Some of these applications are more for novelty, such as Makr Shakr's robotic bartenders, with whimsical names such as "Toni" and "Bruno," that have been making cocktails in bars from London to Dubai over the past few years with more than 2.6 million drinks served!

Niche professions, such as language translators, have come to feel threatened by the ability of machines to translate between multiple languages with startling accuracy, often in real time. Anyone using Zoom videoconferencing during the COVID-19 pandemic might have found quite handy the automatic transcription service provided alongside an automated language translation audio channel for widely spoken global languages. Some analysts believe the future here may be more of a "human on the loop" model, where a human translator uses the results of a machine as a fast first pass and then works to make the translation more accurate and reflective of colloquial language usage. But would we still need as many translators? In a seemingly small industry, will such a specialized skill be needed?

In other domains, such as automated customer support, AI systems that combine speech recognition and natural language understanding— much like those used in smart speakers such as Amazon Alexa and Google Home—have become a new front line for customer interaction. Only when the system isn't able to get you the help you need will a human be brought in to intervene. That seems very likely to reduce the number of customer support representatives needed.

How about the world of finance? Increasingly, AI is used to power ever-more-sophisticated models for trading equities, commodities, and derivatives. Called "quantitative hedge funds," these systems use algorithmic techniques, frequently powered by machine learning models, to make split-second transactions on everything from Apple stock to zinc futures. One of the pioneers in this area, Renaissance Technologies, was founded by Jim Simons, a mathematics PhD who had spent time working at the National Security Agency on code breaking—a somewhat different but equally quantitative form of pattern recognition. Renaissance's Medallion fund has generated more than $100 billion in profits since the company's founding in 1982. Indeed, the world of finance has shifted in the wake of technological developments, as financial services firms increasingly focus on customer relationship management and access to private funds utilizing quantitative technologies. We might say that finance is a field ripe for the adoption of advanced technologies like AI. If markets can be inefficient when filled with slow, error-prone people, why shouldn't AI be used to help optimize pattern identification and exploit inefficiencies far faster than humans can possibly do? Will that lead to fewer humans employed by investment firms? A great deal of ink has been spilled by boosters and naysayers arguing both sides of these questions.

But what about your doctor? Does she face a threat from AI? Jobs we once thought unassailable by technology are now receiving closer scrutiny. Take breast cancer detection. A research team from Google reported on an "artificial intelligence (AI) system that is capable of surpassing human experts in breast cancer prediction." They noted, "In an independent study of six radiologists, the AI system outperformed all of the human [mammogram] readers." Similarly, a team from Stanford developed "an algorithm that can detect pneumonia from chest X-rays at a level exceeding practicing radiologists." Developments such as these led Geoff Hinton, a pioneer in neural networks and deep learning and a winner of the 2018 A. M. Turing Award, to state that "people should stop training radiologists now. It's just completely obvious that within five years deep learning is going to do better than radiologists." That was in 2016.

Since that time, it has been noted that the work radiologists and other medical professionals do is much broader than just interpreting X-rays. Indeed, there's lots of work involved in preparing patients for exams, determining the need to collect other sources of information such as through biopsies, and interpreting those results into a final diagnosis. Perhaps more important, there's the human factor: people want to interact with other people, not just machines, especially when facing life-threatening news. And let's not forget the notion of legal liability: Who should be sued if a machine makes a mistake? With a doctor involved, that question, for better or for worse, is much easier to answer.

The extent to which AI technology will threaten to displace highly skilled workers such as doctors ultimately remains to be seen. A 2019 study in the United Kingdom summarized the state of affairs, stating "Our review found the diagnostic performance of deep learning models to be equivalent to that of health-care professionals," but then went on to conclude that "poor reporting is prevalent in deep learning studies, which limits reliable interpretation of the reported diagnostic accuracy." In other words, the deep learning models may be able to match human performance at the narrow task of making a diagnosis from an X-ray, but the fact that such studies don't then use the results of the algorithm in an actual medical setting means it's not possible to determine whether the model's prediction would have actually led to a better outcome for the patient. They also note that the importance of understanding the reasoning behind the algorithm's diagnoses is "fundamentally incompatible with the black box nature of deep learning, where the algorithm's decisions cannot be inspected or explained."

AI has come a long way since the notion of an intelligent machine emerged in the 1950s. Whereas AI has impacted our lives in mostly invisible ways in the past, we're now poised for a more visible and widespread revolution. Many of its benefits are clear. But what might we lose in the process?

IS AUTOMATION GOOD FOR THE HUMAN RACE?

Automation might be threatening the jobs of blue-collar and white-collar workers alike, but the advance of AI is proving to be a boon for at least one tiny employment category: philosophers. The past decade has witnessed a blossoming of work on ethics and AI. Tech companies look to hire chief ethics officers, nongovernmental organizations are staffing tech ethicists, and think tanks as well as universities frequently advertise new positions at the intersection of ethics and technology. The result, at least to date, is nothing short of a confusing blizzard of AI ethics, principles, guidelines, and frameworks issued by individual companies, NGOs, government committees, and blue-ribbon commissions. According to one recent count, more than eighty distinct AI ethics documents have been released in the past decade, with the overwhelming majority appearing since 2016.

Some of the corporate efforts at AI ethics don't amount to much more than vague platitudes, essentially public relations–driven "ethics washing." In 2019, Google announced the creation of an expert external advisory council—an AI ethics board—to guide and monitor the responsible development of AI by the company. After public controversy arose related to members of the council who opposed rights for LGBTQ people and who sponsored work that questioned human-generated climate change, Google dissolved the council in less than a week and did not attempt to form it anew.

Even in cases where the frameworks are more serious, they often traffic in airy generalities. An ethics framework that states as general principles that an AI system must promote justice and be fair, bias free, privacy protecting, and accountable does not really provide much guidance. As we've seen in earlier chapters, the interesting and relevant issues arise when we ask deeper and harder questions about what fairness means, whether it is possible for an algorithm to encode it, and if it is, what price in predictive power we should be willing to pay to ensure it. Of course privacy is valuable, but an AI principle that announces a firm commitment to it tells us nothing about how we should balance the trade-offs between privacy and security.

In AI, our essential task is to understand how smart machines affect the very possibility of human flourishing. We must examine whether smart machines enhance the capacity of individuals and societies to flourish, and we should strive to identify approaches to developing AI and policies to govern it that will harness its great potential and minimize its considerable risks. There are two key questions: First, under what circumstances do massive advances in AI threaten to undermine human agency and perhaps challenge the very idea of what it is to be human? Second, what does increasing automation do to the material well-being of workers whose jobs are displaced or transformed?

PLUGGING INTO THE EXPERIENCE MACHINE

To understand the interaction between smart machines and the value of human agency, it is useful to examine a hypothetical scenario devised by the philosopher Robert Nozick in 1974. Welcome to "the experience machine."

Imagine you had access to a machine that could provide you whatever pleasurable experiences you desired. These could be as trivial as the experience of riding a roller coaster, eating an ice cream cone, or dancing to your favorite music. Or they could be as profound as falling in love and being loved in return, writing and performing music, or bringing about world peace. If you felt your imagination was impoverished, you could access a library of options drawn from literature, film, and travel. The experience machine gives you the full sensation—the lived and felt reality—of experiencing anything as if from the inside. You could program your wildest dreams as well as your most familiar pleasures. And imagine that you can do so today, tomorrow, next month, or even for the rest of your life. You could plug into this machine and be guaranteed to feel the experience of happiness forever.

To create the experience machine, Nozick imagined using "super-duper neurophysicists." Today, we need not imagine a brain scientist but the actual computer scientists behind virtual reality devices. Instead of the hypothetical example of the experience machine, imagine the Oculus Rift—the powerful VR goggles manufactured by Facebook—on steroids. This is not quite as preposterous or unimaginable as it may sound. Palmer Luckey, a cocreator of Oculus, had in mind a set of gaming experiences in VR, but in interviews he also expressed a far grander aspiration. He spoke of a "moral imperative" to bring VR to the masses so that they, too, and not only the wealthy or geographically privileged, could experience the good things in life such as a sunset over the Aegean Sea, the *Mona Lisa* in the Louvre, the Great Migration on the Serengeti, or a Bruce Springsteen concert in New Jersey. "Everyone wants to have a happy life," he said, and virtual reality "can make it so anyone, anywhere can have these experiences."

If you could plug into a virtual reality experience machine, would you? Should you? Luckey was asked this question by a journalist, and said he would "absolutely" plug in: "If you asked anyone in the virtual reality industry, they would say the same."

Nozick thought it obvious that *no one* would plug into the experience machine. "We care about more than just how things feel to us from the inside; there is more to life than feeling happy," he wrote.

Nozick was posing a simple and fundamental question, familiar not only to philosophers but to everyone who has wondered about our purpose in life: Is happiness, specifically the experience of being happy, the only important thing in life? In answering that question in the negative, he was taking aim at utilitarianism, the philosophical creed first developed by Jeremy Bentham. Utilitarianism holds that the ultimate good in life—the *summum bonum*—is happiness, understood as the experience of pleasure, and that the morally correct action for any individual is that which maximizes happiness for all, the greatest good of the greatest number. The implicit moral task of technologists, then, is to identify the best means to maximize happiness.

In saying that anyone reading about the experience machine would decline to plug in, Nozick was pointing to the significance of knowing that our own agency, our own effort and talent, are connected to the experiences we have in life. "We want to be importantly connected to reality, not to live in a delusion," he said. What's troubling about the experience machine is that apart from deciding to plug in, our own efforts bear no causal connection to our experience. True happiness can be achieved only when we bring about our own pleasure or well-being, not when you get its simulacrum for free.

The most incredible virtual reality device might have some appeal, but would you really want to plug in forever, even if you were guaranteed the maximal flow of pleasure you could imagine? For many people, the answer is no. We generate meaning in life not merely from the felt experience of pleasure, pain, or anything in between but also from knowing that our actions, our very intentions and efforts, imperfect though they may be, are directly and causally connected to what we experience.

One of the pioneers of virtual reality is Jaron Lanier, now one of the most trenchant critics of the development of technology in the past decade. Lanier's 2010 book *You Are Not a Gadget: A Manifesto*, written before Facebook had turned a profit, anticipated many of the privacy-abusing practices so familiar in technology today. It also made the startling claim that certain features of digital technology "tend to pull us into life patterns that gradually degrade the ways that each of us exists as individuals." In other words, we can lose our humanity if we don't attend carefully to what technology is doing to us.

There are some forms of labor we simply don't want to replace with machines if we're at all interested in maintaining a society in which individuals find meaning and motivation in their daily lives. Even if machines can outperform humans at certain tasks, we might decide that we either want to do the task ourselves or craft the technology in a way that preserves human interaction. And given the pace of automation in our lives, we have to attend not just to deciding on the importance of human agency and labor in any particular case but also to the slow and steady outsourcing of our agency to machines, one small increment at a time. The aggregation or accumulation of smart machines in our lives and the death-by-a-thousand-cuts degradation of human agency are the more insidious problems we face.

We don't want to overdramatize the idea of human agency or romanticize the importance of human labor. When it comes to what gives life meaning and what produces individual well-being, there is no shortage of tasks, such as computing spreadsheets, that we should gladly give over to machines. If we can automate forms of labor that are boring, exploitative, alienating, or dangerous, we should. But we might equally seek to develop smart machines that augment rather than displace human agency in the arenas that are essential to our well-being and our very sense of humanity. This involves both the design of technology and the policies that regulate its use. Smart machines might increasingly outperform humans and deliver greater productivity, but at bottom, we can't automate human flourishing.

THE GREAT ESCAPE FROM HUMAN POVERTY

Another element essential to human flourishing is material welfare. Every person must attain a minimum level of material well-being in order to meet basic needs and escape the misery and degradation of poverty, which dooms a person to unhappiness and a predictable assortment of ills. Just as an animal needs sufficient food to eat and a vegetable garden sufficient nutrients in the soil to grow, humans need sufficient resources to enable the capabilities that make a life worth living.

A shocking fact of our story on planet Earth is that desperate poverty has been the condition of most human beings for most of recorded history. According to the economist Gregory Clark, "The average person in the world of 1800 was no better off than the average person of 100,000 B.C." It's only recently, in the aftermath of the Enlightenment and the scientific discoveries and technological innovations that powered the Industrial Revolution, that many humans overcame total impoverishment. Yet desperate poverty, often defined as living on less than two dollars per day, is still the punishing fate of many people today. The Nobel Prize–winning economist Angus Deaton described the past two hundred years of history as "the great escape" and emphasized that rising levels of wealth are connected with improved levels of health. One remarkable indicator of progress: In 1916, the average American male could expect to live 49.6 years and the average American female 54.3 years. By 2016, life expectancy for American men and women was beyond 75 years. What good is human agency if you're no longer alive to exercise it?

We often remember the twentieth century as an era defined by devastating world wars and the creation of atomic weapons so powerful they could destroy Earth many times over. But it was also the era in which hundreds of millions of people finally had material wealth sufficient to meet their basic needs. Recent economic growth in India and China has lifted billions more out of desperate poverty. As nations have grown wealthier, people in rich countries have reported less pain and disability, IQ scores have been rising, and people have been getting taller because of improved nutrition and the massive decline in hunger and famine. Achieving a sufficient level of well-being may not guarantee happiness or flourishing, but it is clearly a precondition for it.

What level of material welfare is needed? Clearly we must move beyond the marker of two dollars per day of desperate poverty. In wealthier countries, income levels must be even higher to meet basic needs of food, clothing, shelter, access to health care, and education. The relevant standard is sufficiency, which stands in contrast to equality. A society that facilitates basic material needs does not have to make everyone equal in income or wealth. But it must guarantee that every person has enough.

What effects might automation have on the ability of every person to achieve this standard? To date, technological innovation has been one of the most important engines of economic growth, lifting much of humanity out of poverty. An age of smart machines that delivers increased efficiency in the workplace might well bring about a new chapter in economic growth and productivity. Of course, there's a huge downside risk as well: if the age of automation displaces huge numbers of people from the workforce, many people will lose a reliable source of income and suffer a threat to their material welfare.

It's been said that if big data is the new oil of the economy, then AI is the electricity. Andrew Ng, a leading AI scientist, says that AI will produce "automation on steroids" and transform every industry known to humankind. The benefits of increasing automation are easy to see, but the costs are often concentrated and sometimes hard to pin down. Some of the costs result in power and wealth inequities. The CEO and shareholders of a ride-hailing service that deploys millions of self-driving cars stand to gain untold wealth, while millions of unemployed taxi, bus, and truck drivers will have to grapple with the consequences of technology over which they have no power. When it comes to material well-being, the age of smart machines may be wonderful for some but livelihood destroying for others.

WHAT IS FREEDOM WORTH TO YOU?

One hidden yet important cost of automation is that it may undermine our freedom to live our lives as we want. Let's return to the example of self-driving cars, which could save up to 300,000 lives *per decade* in the United States alone—something one reporter called potentially the "greatest public-health achievement of the 21st century." But with this transformation in mobility, a number of things would be lost. For one, a lot of us get a great deal of pleasure out of the act of driving: the open road, the kids in the back, the radio on, your hands on the wheel. Getting a driver's license is a veritable rite of passage for teenagers. The freedom to pilot yourself wherever you need to go, at whatever time and whatever speed, is a cherished part of the driving experience. Is it possible for an autonomous vehicle to deliver *Fahrvergnügen*—the wonderful German word for "driving pleasure" popularized in Volkswagen advertisements in the 1990s? Count us as dubious. And we're not alone. Though his Secret Service detail would do it for him, George W. Bush still drives on his ranch in Texas because he enjoys it. Especially so because former presidents of the United States are not allowed to drive on public roads for security reasons. To achieve the full benefits of automation in driving, we'd have to give up some of the agency we enjoy and be willing to sacrifice the pleasure we get from this activity.

How does one attach a value to achieving something through the exercise of our own powers? We are used to thinking about costs in purely material terms—things that can be quantified and counted. This is why income or wealth is so often taken as a proxy for well-being. But as the example of the experience machine suggests, it is important to many of us *how* we achieved that happiness and pleasure and whether in fact we contributed to it through our own effort.

Yet attaching a value to human agency isn't straightforward. This is a challenge Amartya Sen took up in his effort to reimagine the purpose of economic development. He took as his starting point Aristotle's famous dictum that "wealth is clearly not the good we are seeking, since it is merely useful, for getting something else." Sen didn't want to measure only whether countries were rich or poor. He wanted to focus attention on what material wealth brings—whether people in any country have the freedom to live as they would like, to develop and deploy their interests and talents. To Sen, economic development is first and foremost about attaining freedom and unlocking the exercise of human capabilities. This perspective challenges the common conception of well-being. It forces us to consider not only wealth but the key ingredients that enable people to lead a full and meaningful life, such as education; a long life expectancy; access to clean water, food, and health care; political freedom; and civil liberties.

The United Nations Development Programme (UNDP) took Sen's proposal seriously in an effort to challenge the world's reliance on measures of gross domestic product (GDP) as a proxy for well-being. It invented a new measure, the Human Development Index (HDI), to better capture people's capabilities. Although income is part of the HDI, the index also includes years of education and life expectancy. It is still a rough proxy but one that comes much closer to capturing something about the ability of human beings to pursue their own goals and aspirations, to flourish as individuals and societies.

When one takes this idea into the world, dramatic differences in the quality of people's lives quickly become visible. For example, although Equatorial Guinea and Chile have almost the same level of per capita income, the quality of human development is far greater in Chile. Likewise, even at lower levels of income, the differences can be substantial. Because of their commitments to educational opportunity and access to health care, some developing countries, such as Rwanda, Uganda, and Senegal, provide a richer array of opportunities to their residents than do other countries at the same level of income. Though this approach isn't a perfect representation of the value of human agency, it does challenge us to think about a world in which our income remains the same or even increases because of automation but in which we lose something else that's important: flexibility and the freedom to make choices about how we live.

THE COSTS OF ADJUSTMENT

Of course, automation might also diminish people's employment opportunities and income. The fear of this is what drives the headlines, though it is surprisingly challenging to measure in practice. One approach is to identify the particular jobs that are at risk. A group of economists at Oxford University examined the 903 detailed occupational categories coded by the US Department of Labor, evaluating them in terms of the knowledge and skills they require as well as the likelihood that advances in automation could displace the need for human labor. The report estimated that nearly 50 percent of jobs in the United States are at a high risk of technological displacement via automation by 2030. By contrast, economists at the Organisation for Economic Co-operation and Development (OECD) in Paris undertook a similar exercise and estimated that only 9 percent of jobs are truly threatened.

These empirical disagreements mirror the overhyped rhetoric that always seems to accompany a new wave of technological change. In the 1930s, the English economist John Maynard Keynes worried aloud that "we are being afflicted with a new disease of which some readers may not yet have heard the name, but of which they will hear a great deal in the years to come—namely, *technological unemployment.*" Two decades later, the Harvard economist Wassily Leontief opined, "I do not see that the new industries can employ everybody who wants a job." But for everyone worried about mass unemployment, there are those with a more optimistic take. Hal Varian, the current chief economist at Google, assessed the potential employment losses from AI as follows: "If 'displace more jobs' means 'eliminate dull, repetitive, and unpleasant work' the answer would be yes."

There is not one clear answer to the question of how automation will impact the material welfare of workers. Some occupations will be more affected than others. The experts say that dentists and clergy remain relatively safe from AI at this point, while customer service agents, telemarketers, accountants, and real estate agents should be concerned. It's also far more likely that automation will change how most people work, rather than eliminating occupational categories outright. Smart machines can take over the repetitive and dull tasks while leaving human beings to focus on things that require greater cognitive capability or creativity.

For example, it's clear that the effects of a shift to self-driving cars wouldn't be felt equally across society. For some of us, it might be a treat to be picked up by an automated vehicle and whisked off to our next destination. And for some, such as the disabled or elderly or inebriated, low-cost mobility could be a game changer. But for others, driving is more than a way of getting from one place to another; it's an occupation. Nearly 3.5 million Americans are employed as professional truck drivers. An additional 5.25 million work in the trucking business. More than 40 percent of these employees are minorities, and many live in states with lower-than-average incomes and fewer public benefits, including in the Midwest and South. Whether our vegetables and fruits arrive via an automated vehicle or a human-driven truck may not matter to those of us who play no role in its delivery. But that transition would have significant consequences in the workforce, creating a new class of people whose livelihoods have been significantly disrupted.

It's more challenging still to estimate the aggregate effects of automation on the economy. Though the potential displacement of workers is one cost, automation may have beneficial effects on the economy as well. For example, just as the advent of new technology in agriculture led to greater yields from the same fields, automation is likely to make workplaces more productive. A world with greater automation may also be one with new job categories and with specialized skills in high demand. Building this more automated future will require significant capital investment. So all things considered, it may be that automation would improve our total material well-being, even as particular industries or individuals experience significant displacement.

One team of economists set out to capture these overall effects quantitatively using recent data from the United States. They examined changes in the industrial use of robots between 1990 and 2007, looking at the ways in which the adoption of computer-assisted technologies affected overall employment and wages. They found that the growth of robots was associated with a reduction in both employment and wages. By their estimates, each additional robot in a region reduced employment by 6.2 workers—a large but plausible effect. However, because regions do not operate in isolation, the difficult challenge is to figure out how these effects add up across the economy. With the use of robots, the costs of production go down, potentially generating greater economic activity elsewhere in the economy. When one takes into account these countervailing effects, the impact on both employment and wages becomes significantly smaller.

The implication is that beyond the loss of human agency, our primary concern should be how the consequences of automation are distributed. As with the truck drivers whose livelihoods are threatened by self-driving vehicles, the risks of automation are unevenly distributed across occupations and income groups. By one estimate, those earning less than $20 per hour face a nearly 83 percent probability of displacement by automation, while those earning more than $40 per hour are almost immune to the risk. The costs of lost jobs, the impact on families, and the need to learn new skills will be substantial. At this stage, though the technology companies reap the benefits of deepening automation, they bear no formal responsibility for helping workers transition in the face of automation; the costs are dumped on the doorstep of government.

For example, Arizona is ground zero for the testing of autonomous vehicles, and some local citizens are not happy about it. They've slashed tires, hurled rocks at the cars, and attempted to run them off the road. We can see around us the beginning of a modern revival of the Luddites, nineteenth-century textile workers who destroyed machines they believed might one day replace them at their jobs: taxi drivers organizing against ride-share applications in New York and other major cities; Barcelona's crackdown on Airbnb in an effort to sustain the hospitality industry and prevent further urban gentrification. One recent Nobel Prize–winning economist, Paul Romer, even warned that as anger at tech companies boils over, it may result in the bombing of data centers. But like that of the Luddites, whose tactics did not achieve their goal of stopping industrialization, violence aimed at destroying computing machinery will not stop the juggernaut of automation.

SHOULD ANYTHING BE BEYOND THE REACH OF AUTOMATION?

Maybe there are some things we should never automate. Stuart Russell, one of the world's leading computer scientists, a professor at UC Berkeley, and the author of the leading textbook on artificial intelligence used at more than 1,400 universities, thinks we must draw the line in a particular domain. In 2015, he published an open letter on behalf of AI and robotics researchers regarding the use of autonomous weapons.

Autonomous weapons have been described as the "third revolution in warfare, after gunpowder and nuclear arms." These weapons are able to select and engage targets without any human intervention. In contrast to remotely piloted drones or cruise missiles, autonomous weapons require no human involvement in targeting decisions. Militaries using autonomous weapons would be able to identify predefined criteria for a target and then deploy weapons to carry out a mission with humans completely out of the loop.

In his open letter, Russell worried about the potential for a global arms race in AI weapons. "Autonomous weapons are ideal for tasks such as assassinations, destabilizing nations, subduing populations, and selectively killing a particular ethnic group," he warned. Though AI offers substantial benefits for humanity, including the possibility of a safer battlefield, he called for a ban on "offensive autonomous weapons beyond meaningful human control."

Since Russell published his open letter, more than three thousand individuals and organizations have indicated their support for an autonomous weapons ban and signed a pledge not to "support the development, manufacture, trade, or use of lethal autonomous weapons." The signatories include major names in technology, including Elon Musk (SpaceX and Tesla), Jeff Dean (the head of Google AI), and Martha Pollack (the president of Cornell University), and leading organizations, such as Google DeepMind. A campaign to ban killer robots has gone global, and thirty member countries of the United Nations have explicitly endorsed the call for a ban. The United States is not among them. Neither is China nor Russia.

Maybe autonomous weapons are a relatively easy case for strict limits on automation. It is hard to disagree with the idea that machines should not make life-taking decisions. If a life is going to be taken, it is essential that a human being take responsibility for that consequential decision. Yet it isn't too big a leap from autonomous weapons to the life-and-death decisions that self-driving cars will need to make. Again, perhaps the stakes are simply too high to remove human judgment completely; a human always needs to be in the loop.

There are other domains in which the complete loss of human agency is something we might find unacceptable. For example, though machines may be able to optimize school lesson plans, there is a craft to teaching. No AI system can offer comfort to a student in need or a twinkle in the eye when he or she has done something impressive. Music, art, and architecture might be created more quickly or flawlessly by smart machines, but the value we place on accomplishing these works would be irretrievably lost. This is because we gain something from the very fact that we contributed to the outcomes, especially when their production seems almost superhuman.

Of course, even if machines can do something more efficiently, we may want to reserve the right to do it ourselves simply because we derive pleasure or meaning from it: writing a beautiful piece of computer code, solving a complex math problem, hiring a new employee. Recognizing the value we attach to human agency means figuring out both where it is necessary to prioritize human decision-making and where we don't want to rule out human involvement entirely.

WHERE DO HUMANS FIT IN?

One framework for navigating this terrain focuses on augmenting human capabilities, rather than replacing them, with AI providing suggestions to humans, who then make the final decisions. This is known as having a human *in* the loop.

In some cases, we may be willing to give an autonomous system the ability to act directly in the world and have a human provide supervision only as needed. This second model, called human *on* the loop, essentially allows the autonomous system to carry out its work, but the human can intervene or override the AI as necessary. In the context of autonomous vehicles, Tesla's Autopilot system, which can drive a car autonomously in certain road conditions, requires that drivers remain alert and engaged so they can take over driving when hazards arise.

Left to their own devices, it isn't clear that CEOs, investors, and engineers will value the sort of enhancement of human capabilities that Sen's perspective on freedom demands. For someone running a business, the calculation is simple. Current market incentives push the development of AI toward displacing workers. If a shift to automation can improve the bottom line, it would be irresponsible, especially from the perspective of shareholders, not to transition away from human labor. Worse, the tax code in the US exacerbates the situation by taxing labor at rates of roughly 25 percent but equipment and software at less than 5 percent, an effective subsidy to corporations that purchase machines and software that automate the workplace. Clearly other stakeholders are impacted by these choices: the workers themselves, their families, and the governments that will have to deal with the consequences of their unemployment. Where do they fit into the decision-making process?

Iyad Rahwan, a Syrian Australian computer scientist and director of the Max Planck Institute for Human Development in Berlin, has been working for years at the intersection of computer science and society. He believes that we need to move beyond the idea of "humans-in-the-loop" as we confront the potential uses of AI that have widespread societal implications. In this context, when further automation implicates all of us, we need to ensure that society's values are reflected in the ways that machines are programmed and ultimately deployed. He calls this keeping "society-in-the-loop."

Those of us who aren't technologists might just call this politics. Rahwan's argument is that programmers, designers, and executives shouldn't be the only ones balancing competing values, lest society's trust in the advance of artificial intelligence be permanently eroded. This is a powerful idea in principle. But what might it mean concretely as we try to shape a new future of work?

At a high level, it means empowering and amplifying the voices of workers. The long-run erosion of worker power and union organizations has meant that corporate executives are largely free to set priorities, cut costs, and invest in new technologies without due consideration of the potential impacts on their workforces. And in an environment in which more open international trade creates a race to the bottom on standards and worker protections, those who are most susceptible to job loss are even more isolated. Darren Walker, the president of the Ford Foundation, which has made tackling inequality its highest priority, lamented, "Too often, discussions about the future of work center on technology rather than on the people who will be affected by it."

But there is a different path. For example, when the California health care provider Kaiser Permanente rolled out a new electronic health records platform, it was a union that secured job and wage protections, training guarantees, and a channel for workers to shape how the new technology would be used. Thus it's no surprise that this era of automation is also one marked by growing calls for worker organization and a revolution in how corporations are structured. Despite corporate resistance, workers in Amazon warehouses and engineers at Google have already taken the first steps to unionize, recognizing the need to counteract the disproportionate power of company executives. A movement is also afoot to increase the number of worker cooperatives. Owned and governed by employees, these new corporate models distribute the profits from their operations more equitably. In the United States, there are only three hundred or so "democratic" workplaces, according to the US Federation of Worker Cooperatives. But in other parts of the world, including some European countries, they are a common and accepted feature of the corporate environment.

Beyond job and wage protections, what else might workers demand? They might focus on the ways in which automation does more than displace humans. It also transforms the work environment, imposing new risks that must be borne by employees rather than employers. Not long ago, for example, fast-food restaurants, car dealers, customer service phone lines, and stores in a mall had to provide stable and predictable work schedules. This came at a cost to owners, who had to pay for the oversupply of labor in the face of weak demand. The introduction of algorithmic tools to schedule employees aims to deliver the optimal rate of staffing and has improved efficiency from the standpoint of employers. But it has come at a cost to employees, who now must agree to "just-in-time" scheduling practices and greater uncertainty and instability about shifts and total hours worked, making it difficult for them to meet some of their basic human needs.

Business leaders are also beginning to rethink the proper role of a corporation. In 2019, the Business Roundtable, a clubby network of powerful corporate elites, released a new "Statement on the Purpose of a Corporation" that was signed by 181 CEOs. Though the statement reaffirmed the principle that shareholder interests are primary, it also embraced a broader vision of company responsibility that includes investing in employees, dealing fairly and ethically with suppliers, and supporting the communities in which they work. But without significant changes in public policy, these principles end up being hollow promises.

Senator Elizabeth Warren of Massachusetts made a push for "accountable capitalism" a hallmark of her pro-worker agenda during her run for president in 2020. From her perspective, the transformation of corporate America requires a new obligation for company directors to consider the interests of all stakeholders, not just shareholders; significant representation of workers on companies' boards of directors; restrictions on how corporate equity is handed out to reduce incentives for short-termism; and greater worker voice on how corporations use their political influence and spending. None of these ideas is truly revolutionary, except in the United States. They mostly reflect a European model of capitalism, empowering workers in addition to owners to chart a company's path.

Policy makers can also put a thumb on the scale for AI development that augments rather than undermines human capability, for there are many ways in which AI technologies can be developed, with widely varying implications. If we think of AI as an alternative to human labor and cognition, the costs to our agency and well-being will be dramatic. But if political leaders use research-and-development investments and tax incentives to drive AI in ways that create new high-productivity tasks for human labor, it could become a new engine of productivity and growth. In education, this could involve a revolution in personalized learning that changes the teaching profession and improves student outcomes. In health care, it could mean moving past a focus on replacing radiologists with X-ray-reading robots to the design of AI applications that empower health care providers to offer more real-time health advice, diagnosis, and treatment.

WHAT CAN WE OFFER THOSE WHO ARE LEFT BEHIND?

The third piece of the puzzle is what to do about the impending distributional consequences of automation. We need to attend to those who might temporarily or permanently lose their chief source of material well-being.

When Andrew Yang launched his candidacy for the presidency in 2019, he was a total unknown. Only forty-four years of age, he had worked in the business world and founded Venture for America, a fellowship program that placed recent college graduates in start-ups to prepare for an entrepreneurial career. He was the first Asian American to run for the presidential nomination of a major party, and though at first his poll numbers were miserable, his campaign managed to capture the imagination of young people in particular.

A single idea defined his policy platform: the freedom dividend, a universal basic income (UBI) that would give every American adult $1,000 per month, regardless of his or her work status. Though the idea has a long history among intellectuals and policy makers, his version of the plan was uniquely tailored to the AI moment. As he explained, "The big trap that America is in right now is that as artificial intelligence and autonomous cars and trucks take off, we're going to see more and more work disappear and we're not going to have new revenue to account for it." His view was that people need an unconditional basic income as a cushion from these systemic shocks, and the companies that will most benefit from automation—the tech companies—should be the ones to pay for it. "So the way we pay for a universal basic income," he argued, "is by passing a value added tax which would get the American public a slice of every Amazon transaction and Google search."

You might think that business leaders would oppose Yang's proposed solution. But Bill Gates has also proposed the notion of a "robot tax," where companies that displace workers through machines would be taxed on those machines in a way similar to the human worker. In Gates's words, "Right now, the human worker who does, say, $50,000 worth of work in a factory, that income is taxed and you get income tax, social security tax, all those things. If a robot comes in to do the same thing, you'd think that we'd tax the robot at a similar level." He went on to argue that those funds could be used "to at least temporarily slow the spread of automation and to fund other types of employment." There's no reason that the funds couldn't also be used to support a universal basic income.

How to pay for UBI is an important issue because the potential cost is enormous. The Center on Budget and Policy Priorities, a nonpartisan think tank, estimated that a basic income of $10,000 per year in the United States—$2,000 less than Yang's proposal—would cost the government $3 trillion per year. By contrast, the US government's largest social welfare program currently, Social Security, cost $988 billion in fiscal year 2018.

In advancing a vision of a universal basic income, Yang was in august company. The list of UBI champions includes tech titans, such as Mark Zuckerberg and Jack Dorsey; champions of progressive policy reforms, including civil rights activist Dr. Martin Luther King Jr., former president of the Service Employees International Union Andrew Stern, and Ai-jen Poo, executive director of the National Domestic Workers Alliance; distinguished politicians, including former Treasury secretaries George Shultz and James Baker; and conservative economists, such as Milton Friedman and Martin Feldstein.

Yet as a policy antidote to mitigate the harms of automation, the push for UBI seems a bit misplaced. As Jason Furman, the chairman of President Obama's Council of Economic Advisers, noted in a speech in 2016, "The issue is not that automation will render the vast majority of the population unemployable. Instead, it is that workers will either lack the skills or the ability to successfully match with the good, high-paying jobs created by automation." His conclusion is that we should not be planning for a world in which people find themselves permanently unemployed; we should focus instead on helping households navigate the dislocation caused by automation and fostering the skills, training, and other assistance needed to get people into productive, high-paying jobs.

If the push for UBI reflects a willingness to make larger investments in dealing with the consequences of automation, there are other ways to spend public resources. The US government could substantially increase the generosity of its social safety net to help the particularly vulnerable manage the temporary loss of employment with generous unemployment benefits, nutritional assistance, affordable health care, and subsidized child care. It could also make significant investments in education and retraining to make it possible for today's employees and the next generation to thrive in an increasingly automated economy. Given that the social costs of automation are a by-product of corporate decisions, it makes sense to think that companies should play a central role in helping government manage these consequences. Gates's robot tax is one way of supporting substantial new spending, as are European proposals to tax digital services. Other ideas that go beyond taxation are actively under debate—for example, the concept that companies should make a binding commitment to donate any windfall profits from AI to address the consequences of automation.

To some observers, arguments such as these raise concerns about big, inefficient, and paternalistic government programs. But setting aside ideology for the moment, there are highly effective social safety nets in Europe that keep people out of poverty and foster greater opportunity and upward social mobility. Most Americans are unaware of how powerful they have been in practice.

In terms of government resources alone, European countries make far larger investments in social spending. In 2019, the United States spent about 18.7 percent of GDP on social programs, whereas the EU average was above 25 percent. The more important story is that once tax breaks are included, the United States actually devotes far more resources to social spending, including pensions, health care, family support, unemployment, housing assistance, and other benefits. The additional spending comes as a result of compensating the private sector with tax breaks when it provides certain social benefits, such as health insurance and pension contributions. And when you add in what Americans spend privately on these social programs, the United States actually spends more than most other advanced economies do. It's just that its spending is far less effective.

Health care is the most obvious culprit: Americans spend an awful lot on health care and get far less in terms of positive results. In most countries, outcomes such as premature deaths among both men and women go down with increases in spending. The United States is an outlier: its rate of premature deaths is identical to that of countries that spend one-tenth the amount on health care. The same is true of infant mortality.

The outcomes are just as disappointing when it comes to social mobility. The idea that Americans can move up the income ladder is central to the country's identity. The reality is that Europeans born in the bottom 20 percent of the income distribution are almost twice as likely to move to the top 20 percent as Americans are. According to Harvard economist Raj Chetty, "you'd have a better chance of achieving 'the American dream' if you're growing up in Canada or many Scandinavian countries than the United States." This is what effective social spending can buy.

When it comes to navigating the dislocation that automation is likely to bring, the US approach will not cut it. A universal basic income is not likely to do the job, either. We'll need an all-hands-on-deck approach to mitigating the consequences of temporary unemployment and building an education and training system that will prepare people for the jobs of the future. This doesn't require radical new policy mechanisms; what's needed is a fulsome embrace of social safety nets. And although this may cost more than what is currently spent in the United States, as the Europeans have shown, it's not only how much you spend but how you spend it that makes all the difference.

●　■　●　■　●

Because Silicon Valley is an epicenter of the design and testing of self-driving cars, it is not uncommon to pull up to a stoplight and see an autonomous vehicle operating right beside you, its sensors visibly sitting atop the roof of the car. For most people, the sight of a self-driving car causes excitement and perhaps spurs the imagination. But when we stop to consider the vast amount of automation that awaits us, it is difficult to look away from the important decisions we will be confronting.

Changes in technology will transform how we work and how we live. Some people will excel in this transition; others will be left behind. Ultimately, however, the way in which automation affects our society is not preordained. It will depend on the choices we make about where and when to retain human judgment, whether to prioritize the augmentation of human capabilities versus their replacements, and how best to support one another as some roles and occupations disappear forever, while others that we cannot yet imagine take their place.

CHAPTER 7

WILL FREE SPEECH SURVIVE THE INTERNET?

On January 6, 2021, while vacationing on a private island in French Polynesia, Twitter CEO Jack Dorsey received an urgent call from his top lieutenants. A decision had to be made about whether President Donald Trump should be temporarily suspended from the platform. Trump's supporters, stoked with misinformation about election fraud, had stormed the US Capitol. The executive team decided to place a twelve-hour suspension on his account. But after the initial suspension was lifted, the soon-to-be-former president continued to post inflammatory rhetoric. On January 8, Twitter ultimately decided to permanently suspend Trump's account, removing it from public view. In a post justifying the decision, it noted that "the President's statements can be mobilized by different audiences, including to incite violence." Dorsey later further defended the decision, writing that "Offline harm as a result of online speech is demonstrably real, and what drives our policy and enforcement above all."

The decision to deplatform the sitting president of the United States brought swift reactions worldwide. Between declarations that it was a long-overdue action to stop a fountain of dangerous misinformation and heated cries by others of censorship and left-wing bias stands the central issue of how social media platforms actually deal with the millions of daily posts containing hate speech, misinformation, and disinformation. Twitter's choice to suspend Trump's account was just the most prominent and consequential attempt to respond to a problem it had grappled with for years. It's a problem that affects far more people than just prominent politicians or attention-seeking celebrities. Every single day the large social networks—Twitter, Facebook, Instagram and WhatsApp (both owned by Facebook), YouTube (owned by Google), Snapchat, and TikTok—must decide what text, audio, images, and video are permissible to post and share with others. And sometimes those decisions defy expectations.

In the fall of 2017, as the #MeToo movement was gaining traction, a female writer for Samantha Bee's late-night talk show named Nicole Silverberg posted online a list of the different ways in which "men need to do better." She was quickly deluged with misogynistic and hateful comments. She shared some of this bile on her Facebook page so her friends could see the horrible invective directed at her. Many people posted in the comments section beneath Silverberg's post. One comic, Marcia Belsky, wrote, "Men are scum."

The women were shocked when Facebook deleted the comment and booted Belsky off the platform for thirty days. Outraged that it was evidently permissible to post hateful comments about women but not to post "Men are scum," Belsky found her way to a large Facebook group of female comics. She discovered that several of them also had their posts removed for making similar remarks, such as "Men are pigs" and "Men are trash." They decided to act collectively, posting dozens of times on the same evening the comment "Men are scum." Sure enough, Facebook immediately deleted every single one. The sentence clearly must have run afoul of Facebook's community standards. Is "Men are scum" hate speech? Is directing the comment "Men are trash" to a particular man a form of bullying? Is it hate speech or bullying in some contexts but not in others? These are the questions that the big tech platforms confront endlessly. The decisions they make can affect billions of people.

For Facebook, with more than 2.8 billion active users, Mark Zuckerberg is the effective governor of the informational environment of a population nearly double the size of China, the largest country in the world. Zuckerberg had it right when he said, "In a lot of ways Facebook is more like a government than a traditional company." But it's not a democracy. Zuckerberg is the king or, depending on your point of view, the despotic dictator of a nondemocratic Facebook state. After all, the company is a private entity governed neither by the First Amendment nor by any universal declaration of a right to free expression.

Facebook's control over content is a fearsome power. Critics of one stripe complain that the big tech companies exercise too much control and ban or delete too much. They thwart the ideal of free speech, cherished in liberal democratic societies and central to various declarations of human rights. Of course, the tech companies are criticized from the other direction, too: they take down too little content, willfully permit hate speech and lies, and are unable to control misinformation, disinformation, and other harmful content. Worse, their algorithms sometimes seem to amplify extreme content, promoting virality over veracity.

A clear example surfaced in the 2019 terrorist attack on mosques in Christchurch, New Zealand, which killed fifty-one Muslims and was orchestrated deliberately as a livestream event by the attacker. The horrific video immediately went viral across multiple platforms. Facebook reported that the livestream video had been viewed two hundred times during the live broadcast and four thousand times in total before it was deleted, but that various individuals had tried to re-upload the video more than 1.5 million times in just twenty-four hours after the attack. In the immediate wake of the attack, Facebook suspended all livestreaming of video on its platform.

There was never any question that the video violated the content standards of the tech companies on whose sites it was posted, but they did not have the internal capacity to prevent it from going viral and thereby provided sustenance to a growing global community of white supremacists. Even so, the Christchurch video makes the case for removal easy. In less extreme cases, the questions surrounding content moderation quickly become more difficult to answer.

Hate speech that gives succor to extremists is protected speech in the United States. Nazi sympathizers are allowed to march in Skokie, and so are white nationalists in Charlottesville. But many people want hate speech banned or diminished from the online world. What about more ordinary forms of political communication? Should tech platforms take down doctored videos of political opponents, as was argued in the 2020 US presidential race? What about fact-free speculation about voting irregularities or fraud? An investigation by ProPublica in 2020 found that although Facebook has policies that oppose voter suppression, misinformation that might suppress voting flourishes on its platform, with posts concerning fraud via mail-in ballots and conspiracy theories about stolen elections gaining wide traction. Yet one person's disinformation campaign is another person's attack ad, and nowhere is free speech more sacred than in the political realm.

The upshot is that we now face a trilemma, a tension among three important values. The first is the value of free expression, the individual right to speak and be heard without censorship, which brings with it the benefits of a broad and diverse marketplace of ideas. Yet in the digital age a strong commitment to free speech puts a second value—democracy itself—at risk. The decision by tech companies to permit nearly unconstrained user-generated content in the name of freedom of expression has yielded interference in elections in many democratic societies by hostile foreigners and standing misinformation and disinformation campaigns on issues of all stripes. Should the political posts in the 2016 US election by Macedonian teenagers and Russia's Internet Research Agency be protected by free speech norms? Should it matter if the identical material is circulated by US citizens rather than foreign agents? Should the speech of elected leaders or candidates for public office be treated differently than the speech of ordinary citizens? Just as important, we must confront the algorithmic sorting of users into filter bubbles that contribute to growing polarization, extremism, and decreasing social trust, all of which threaten the health of democracy.

An unwavering commitment to free speech in the digital age can thwart a third value: individual dignity. The internet is what made possible the situation faced by Nicole Silverberg, an online deluge of misogynistic and hate speech. It also makes possible trolling, bullying, and doxxing (maliciously posting private information about people on the internet, such as their phone number and home address), all of which threaten the dignity of individuals. In the summer of 2020, an independent auditing team that consulted with more than one hundred civil rights organizations, undertaken at Facebook's own initiative, faulted the company for elevating free expression over nondiscrimination and hate speech limits. Citing the audit, twenty state attorneys general wrote an open letter to Facebook's leadership requesting that the company take new steps to prevent the spread of dangerous and potentially harmful speech and "to provide redress for users who fall victim to intimidation and harassment, including violence and digital abuse." Paradoxically, the state attorneys general would likely be obligated under the First Amendment to protect such speech, apart from any that incited violence, if it took place in a public park or during a protest.

To make sense of all this and determine what to do about it, we have to understand the vast difference between today's digital speech environment and the pre-digital environment that preceded it.

THE SUPERABUNDANCE OF SPEECH AND ITS CONSEQUENCES

In 1992, a first-year Stanford law student named Keith Rabois walked up to the residence of Dennis Matthies, a lecturer and resident fellow, and yelled, "Faggot! Hope you die of AIDS! Can't wait till you die, faggot!" The offensive slurs were overheard by many people, and the incident was reported to university administrators and detailed in the student newspaper. You can imagine the firestorm of controversy that followed. Numerous student groups and campus leaders condemned Rabois, and some sought to have him penalized under the university's hate speech code. He soon revealed that he had deliberately sought attention, writing in the student newspaper that though his words were about Dennis Matthies, they were not directed at him. Rabois intended, he said, to "expose these freshman ears to very offensive speech" in the hope that others would feel emboldened to challenge what he saw as a stifling multicultural political correctness. The university leadership ultimately declined to take any action, concluding, "This vicious tirade is protected speech." Rabois ended up leaving Stanford to finish his law degree at Harvard. Today he is a prominent venture capitalist.

Were Rabois to go online now and make the identical statements on social media, it's highly unlikely that his words would be treated as protected speech. One reason is that although in the state of California, where he was a student, private universities are bound by the First Amendment, the neighboring campuses of Facebook, Google, YouTube, and Twitter are not. Private companies can create speech codes and limit users' expression in a great variety of ways.

Beyond the legal implications of the First Amendment lie something more important, however: the online world has fundamentally transformed the very nature of free expression.

Whereas once it was difficult to disseminate speech, images, and video, the internet has made it trivially easy to do so. In 1992, Rabois's homophobic rant could reach no more than the small community of people who happened to be standing within earshot. If any other crank, eccentric, proselytizer, activist, campaigner, or advertiser wanted to communicate an outrageous point of view to the world, freedom of speech would permit anyone to stand on the street corner and bark at passersby. Those who had money could publish a pamphlet, send postal mail, or take out an ad in a newspaper, on television, or on radio. Even if he or she had the money to pay for an advertisement, no outlet would be required to run it. The relative difficulty in disseminating ideas and opinions is one of the main rationales for the First Amendment and the value of freedom of expression. If the government has the power to define what is permissible speech, then it can play the role of censor and block the transmission of ideas to wider audiences.

But the internet and social media are changing everything. Fast-forward to 2020, and say you would like to communicate the patently false view that Bill Gates's philanthropic support for a COVID-19 vaccine allows him to implant microchips in everyone who receives the vaccine. Or you want to repeat baseless claims about a "stolen" election. All you have to do is take to Twitter and broadcast to everyone who follows you. Or you can upload an easy-to-produce conspiracy video to YouTube and watch the views rack up. If you're slightly more sophisticated, you can create bots that will post the material on social media networks thousands of times over. In the meantime, others will see your posts and share them with their own networks, creating viral distribution to still more people. The platforms tend to amplify your outrageous comments because their algorithms are optimized for user engagement, and it turns out that many users like to engage with outrageous content.

These are not hypothetical examples. In the wake of the 2016 US election, BuzzFeed News reported that the top fake election news stories "generated more total engagement on Facebook than top election stories from 19 major news outlets combined." During the pandemic, one post on Facebook about the Gates vaccine conspiracy theory was shared more than forty thousand times in just a few days. A May 2020 poll found that 28 percent of Americans, and more than 44 percent of Republicans, believe the story about Gates. No wonder resistance to vaccination is rising. When Gates was asked how he explains the proliferation of such views, he did not hesitate to blame social media, which he described as a "poison chalice," and called for policy makers to act. "I personally believe government should not allow those types of lies or fraud or child pornography."

Whether it's Rabois-style hate speech or conspiracy-driven disinformation, the internet allows anyone to have a global audience. In 2018, Mark Zuckerberg said that his motivation for creating Facebook was to expand the power of people around the world. "Many of us got into technology because we believe it can be a democratizing force for putting power in people's hands. I've always cared about this and that's why the first words of our mission have always been 'give people the power.'" Another way to describe the situation, however, is that everyone with an internet connection can pump whatever they like into the digital public sphere, leading to an explosion of hate speech, misinformation, hoaxes, and crackpot conspiracy theories alongside all kinds of genuine creativity and expression. Zuckerberg said that the initial attraction to technology as a democratizing force was overly optimistic. "One of the most painful lessons I've learned," he confessed, "is that when you connect two billion people, you will see all the beauty and ugliness of humanity."

The best way to describe the new situation is that the internet and social media have created a superabundance of speech: texts, posts, audio, images, and videos. Billions of pieces of new content are posted *every day* on Facebook, Instagram, and WhatsApp. YouTube brags about its "over 300 hours of video uploaded every minute." And Twitter says that more than 500 million tweets are posted on a typical day.

In conditions of superabundance, we need help finding and filtering the information we want to see. Whereas once we simply subscribed to a newspaper or two, watched one of the small number of evening news shows, or asked a librarian to help us find a book, now we need search engines to highlight the most relevant results and algorithms so that we can figure out what we might want to pursue on social media. Just at the moment when the volume of content has exploded, we've lost the traditional curators of our information ecosystem.

It's no wonder that issues of free speech are now being deliberated more by tech companies than by the editorial staffs of newspapers, producers of television shows, or even governments. Online every individual is a publisher. And there are virtually no editors. The internet has democratized voice by removing traditional gatekeepers—including those oriented toward professional norms of seeking truth, checking facts, and respecting expertise—from standing in the way of anyone who wishes to exercise free expression.

This dramatic development has some undeniable benefits: the ability to learn about virtually any topic, share photos and videos with your friends, find people with shared interests over great distances, and experience a greater diversity of perspectives available to anyone who might seek them out. But the flip side of access for anyone with an internet connection is the capacity to direct harmful speech to a potentially global audience through a blog or viral social media post. Open exchange democratizes self-expression, yet the absence of gatekeepers allows speech that corrodes the informational health of democracy or violates the dignity of some people. And greater quantities of online disinformation and hate speech can lead to greater quantities of offline harm to people.

To see the importance of this shift, think back to 1768, when the *Encyclopaedia Britannica* was first published in Edinburgh, Scotland, and soon went on to be heralded in the English-speaking world as a trusted source of information. Eminent writers and scholars were enlisted as contributors to the several thousand pages that came to make up the work. At the start of the twentieth century, *Britannica* was—ironically, given its name—acquired by an American firm and continued to grow in eminence, with contributing authors including Nobel Prize winners such as Milton Friedman and Albert Einstein. With the advent of the internet, it was only natural that *Britannica* would go online. In 2012, the print version was discontinued, and now only digital versions are sold.

Presaging the demise of *Britannica's* print run, in January 2001 Jimmy Wales and Larry Sanger, inspired by the open-source software movement, created Wikipedia, an organically grown online encyclopedia. Wikipedia was meant to provide a free, up-to-date source of information built by the user community and—in contrast to sources such as *Britannica*—free from the control of a central authority. That freedom gave Wikipedia an unprecedented ability to grow quickly. By 2017, the English version of Wikipedia alone contained more than 3.9 million articles in contrast to the roughly 120,000 articles in *Britannica*. The English version of Wikipedia now has more than 6.2 million articles and attracts more than 8 billion page views per month globally. And it's not from people going straight to Wikipedia for information; two-thirds of visits to Wikipedia originate from a query on a search engine such as Google.

Of course, the world of online information vastly extends beyond Wikipedia. It's just the poster child for the fact that any individual, not just experts or editors, can publish information on the Web, with search engines providing the means for mass distribution of that information. That's led to two crucial and intertwined phenomena. First, the power of editorial control—including expertise, investigation, and fact-checking, among other aspects—has moved into the hands of the masses. This has created a great wealth of information, as evidenced by the thirtyfold increase in the number of articles in Wikipedia versus *Britannica* in just a few years. But it also means that such information is much less reliable. Second, as search engines and social networks become primary venues for the dissemination of information on the internet, their power to up-rank, down-rank, and filter content gives a small number of companies virtually unparalleled power to determine what information users actually see in their search results and news feeds.

Now add to that the industry of search engine optimization, or SEO, which exists to help websites become more visible in search engine rankings. Put less charitably—but perhaps more honestly—search engine optimizers seek to manipulate the results that get high rankings through all manner of tactics, some of which are downright nefarious. In the early days of Web search, you could find web pages that had all the words in a dictionary listed at the bottom of the page. The words were in a tiny white font on a white background, so a human wouldn't see them. But they'd be picked up by the search engine, so that the given web page could be a potential match to *any* query in the same language as the dictionary. Such techniques, which are known as "Web spam" and are akin to email spam, seek to get information in front of users, often with the clear goal of having the user read a particular piece of content or buy a particular product. SEO is big business. It's been estimated that spending on SEO will reach almost $80 billion in 2020. Dealing with Web spam is an ongoing arms race between search engine companies and some less-than-reputable search engine optimizers, who use all sorts of techniques to manipulate search engine rankings, including creating thousands of fake web pages that link to the pages they want to promote; buying up once-popular but subsequently defunct Web domains to create new links; and reverse engineering search-ranking algorithms in order to figure out the factors that cause a page to be ranked highly.

Prior to the advent of the Web, people would often seek out information from trusted sources. We would decide what information we wanted and then make an effort to get—to "pull"—what we were seeking. With social networking and other content platforms, we've accepted a "push" model for getting information. When we go on Facebook, Twitter, or TikTok, it's not often that we're looking for something specific. Rather, we're looking to be shown what's "interesting"—what our friends are doing, reading, watching, or saying. Information is being "pushed" at us, and we have little say in or understanding of how that information is chosen for us as we scroll through an endless list of postings. Think about how information consumption habits might have changed from reading entire sections of a newspaper—which could present point/counterpoint articles or op-eds from different points of view—to now just taking the recommendations given to us. Of course, it's not just Facebook; YouTube recommends the next video you should watch, Amazon the next product you should buy, DoorDash the next meal you should eat. We've gone from being active information seekers to passive information consumers. In the guise of helping us find things we like, platforms now play the role of determining what we *should* like—and we often consume those recommendations without considering or even knowing what the alternatives might be.

And there's a feedback cycle at work. If you've read articles that touted the importance of the Second Amendment in protecting your personal liberty, in the future you're likely to see not just articles about guns but those that specifically promote gun rights. Of course, the search engine or social media platform that fed you the article is also selling personal advertising, so if you read many articles about gun rights, you are likely to see ads for gun products as well. The bottom line is that if a platform is trying to maximize engagement or clicks, it won't be optimizing to present you with different opinions or worldviews. It will optimize to give you more of what you're likely to click on—ostensibly, information that is similar to or reinforces what you previously read. There might be a superabundance of speech, but you may be getting just the sliver you're likely to agree with.

WHEN FREE SPEECH COLLIDES WITH DEMOCRACY AND DIGNITY

Freedom of speech is a hallowed ideal in most societies. A vigorous defense of freedom of expression can be found in every major declaration of human rights. Free speech has long been taken to be essential both for a well-functioning democracy and for the freedom and flourishing of all individuals. "If a nation expects to be ignorant and free," wrote Thomas Jefferson, ". . . it expects what never was and never will be." Freedom of speech supports democracy because it produces a diverse marketplace of ideas, helping inform the citizenry and providing the conditions for discussion and deliberation that are needed for making good political decisions. And it is essential for human flourishing because it enables each person to answer one of the most fundamental questions of life, "How ought I to live?" A society that blocks freedom of expression undermines the freedom of every person to live according to his or her own best lights.

One of the most famous defenders of freedom of speech is the nineteenth-century English philosopher John Stuart Mill, whose writings gave birth to the now-familiar metaphor of a marketplace of ideas. In his 1859 book *On Liberty*, Mill offered multiple rousing arguments on behalf of free expression. He believed that the quest for truth and fact, the enlightenment of individuals and society, holding the powerful to account, and the cultivation of individual autonomy all depended on a robust commitment to free speech. We are better off, he believed, when we permit individuals to breathe their opinions and propositions in a common air with a resulting collision and combustion of ideas. Suppression of expression, he felt, is nothing less than the asphyxiation of the individual and the suffocation of society.

Freedom of expression is valuable whatever the truth of our own view. No one has a guarantee of infallibility, so said Mill; "To refuse a hearing to an opinion, because they are sure that it is false, is to assume that *their* certainty is the same thing as *absolute* certainty." Exposing ourselves to diverse and controversial views is necessary for scientific progress, the discovery of new knowledge, and the establishment of fact. This diversity in turn permits all individuals to exercise their freedom to choose their own best paths in life. "He who lets the world," said Mill, "choose his plan of life for him, has no need of any other faculty than the ape-like one of imitation." Denying freedom of speech amounts to nothing less than "robbing the human race; posterity as well as the existing generation."

Mill's defense of free speech is magnificent. But in the age of the internet, an unbridled adherence to it undermines rather than supports democracy. When speech is superabundant, we have a marketplace of ideas of unprecedented size, rife with potentially anonymous speakers and even bots that impersonate real users and propagate speech in record volumes. The result is, as we all know, a marketplace full of toxic misinformation and disinformation.

A fervent believer in the marketplace of ideas, like Mill, would find the volume of speech and the existence of anonymous and bot speakers to be irrelevant. Listeners, after all, have the benefit of more ideas to confront and can test out the content of any idea independently of knowing the identity of the speaker. In essence, the proper response to bad speech is not to restrict it but to counter it with good or better speech. But we have little evidence that our digital public square is leading to truth and knowledge. Did the internet help correct the lie that Barack Obama was not born in Kenya? Did it lead to a factual refutation of the Gates vaccine conspiracy? To the contrary, the current infrastructure of social media has amplified the reach of these messages and increased the numbers of people who were exposed to and then perhaps believed such falsehoods. It also provides disproportionate power to even small groups of individuals who want to pollute the information ecosystem with lies and misrepresentations.

The fact is, in the age of Facebook, Twitter, and YouTube, the speed with which viral ideas travel makes correcting falsity and propaganda extremely difficult. Take, for example, the arguments of Supreme Court Justice Louis Brandeis, who wrote in 1927, "If there be time to expose through discussion the falsehood and fallacies, to avert the evil by the processes of education, the remedy to be applied is more speech, not enforced silence." But in the age of the internet, a quip often attributed (perhaps apocryphally) to Winston Churchill is more likely apropos: "A lie can travel halfway around the world before the truth can get its boots on."

The success of any democratic society requires a healthy information ecosystem of educated and informed voters, opportunities for reasoned exchange and deliberation, the ability to separate fact from fiction, and trust between fellow citizens and between citizens and their leaders. We have ample reason to worry about whether freedom of speech in the internet age supports these conditions. In fact, we have reason to believe that the traditional commitment to freedom of speech in democracies is undermining them through the toxic externalities of the large private platforms. A well-functioning marketplace of ideas would tend to self-correct. But as former president Obama concluded in a 2020 interview, the inability to sort truth from falsity in the online world constitutes the "single biggest threat to democracy" and nothing less than an "epistemological crisis" in which experts are no longer trusted.

If a fervent commitment to free speech is threatening democracy by allowing misinformation and disinformation in our public square, it is also threatening individual dignity. When anyone can upload content to the internet and direct it at specific people or groups, we have a recipe for speech that threatens individuals. The most common and worrisome form of such speech is hate speech. But online content that undermines the dignity of other human beings comes in many other forms as well: trolling, bullying, doxxing, revenge porn, fighting words, incitement to violence, and child endangerment. Speech that harms the dignity of others comes in two forms: that which causes psychological damage and that which can lead to actual violence. Of course, hate speech existed prior to the internet, but online speech raises distinctive and potent challenges because it can be cloaked in anonymity, can be disseminated by bots, and is amplified by the algorithms of the unaccountable platforms. As a result, the harms of online speech are much more difficult to contain.

Now let's acknowledge that even the most die-hard defender of free speech believes that there are some limits to what can be said. When speech, images, or video can be shown to lead to physical violence or imminent threat of harm, speech must give way. In the classic but very over-simplified example, you can't yell "Fire!" in a crowded theater because of the likelihood that physical harm will result as people flee for the exits. Similarly, the Supreme Court permits limits on speech in the case of incitement to violence and "fighting words"—speech that by its very utterance inflicts injury or causes an immediate breach of the peace. Much of this is in keeping with the vision provided by Mill: your liberties guarantee you a wide zone of free expression without interference by others, especially the government, but they do not permit you to harm others.

Using the freedom to create speech codes that is accorded to them as private entities, the large tech platforms strive to limit certain kinds of dignity-violating speech. Facebook provides quarterly data on its efforts to enforce its community standards for content posted on Facebook and Instagram. Twitter and YouTube provide similar reports. Using a combination of automated detection, human moderation, and user reports, all of the major platforms try to remove pornographic content, child-endangering content, bullying and harassment, and content from terrorists and organized hate groups. Though legally permissible, adult nudity and sexual activity are some of the most commonly removed forms of content. In 2019, for instance, Facebook took down more than 120 million pieces of such content.

Where does this leave hate speech? The Christchurch terrorist spent many years in online forums seeking community with white supremacists. But prior to conducting his massacre, he uploaded a manifesto on 8chan, a platform known to be populated by extremists due to its virtually nonexistent content moderation, that stated, "Well lads, it's time to stop shitposting and time to make a real life effort post." Whenever hate speech leads to—and especially if it causes—physical harm, there is a strong case for limiting free speech. The difficulty is how to determine whether the speech was the cause of real-world acts of violence.

This is a demanding standard, so advocates of laws against hate speech often justify restrictions in cases not only of bodily harm but also of emotional or psychological harm. Hate speech that is targeted at people on the basis of group characteristics, such as race, sex, or national origin, or that compares humans to animals, is dehumanizing. It can cause distress and trauma and be an affront to individual dignity.

In recent years, the large platforms have focused attention on bullying, harassment, and hate speech. Facebook reports that though its efforts to proactively identify hate speech are improving, with automated tools removing more than 50 million pieces of content in the first half of 2020, the problem is stubborn because hate speech is so varied, contextual, and controversial. In late 2020, the company estimated that hate speech constituted 7 or 8 out of every 10,000 pieces of content viewed by users. Rather than relying on the voluntary efforts of the big tech companies, should democratic governments impose rules and standards on the companies?

Democratic societies strive to protect individual dignity, as evidenced by human rights documents that often begin with a declaration of the dignity of all persons. Such protections account for the limits placed on free expression in the form of hate speech laws in many different countries. Notably, the United States is an outlier with a virtual absence of hate speech regulation.

So when Keith Rabois uttered his hateful words and repeated them in print, they were permitted under free speech protections. Should the situation be different if the same words were posted online? The internet and social media not only provide a megaphone for hate, they provide the infrastructure to spew hate anonymously through bots and fake accounts. The result is an online environment seemingly tailored for costless expressions of hate. And we all know how invective can be directed at particular people online: flaming, doxxing, and constant harassment meant to drive them off platforms. When this happens, the ideal of the internet as a marketplace of ideas deteriorates into an environment that fails to respect the equal communicative liberty and dignity of all.

WHAT ARE THE OFFLINE HARMS OF ONLINE SPEECH?

Back in 2001, Harvard professor Cass Sunstein warned of the harmful effects of social media on democracy and human dignity. His main concern was that online spaces tend to favor "enclave" deliberation—conversations among like-minded people who encounter information and arguments that reinforce, rather than challenge, their preexisting views. Sunstein argued that, though enclave deliberation is not inherently dangerous, it tends to exacerbate group polarization in practice, breeding extremism and putting "social stability at risk." Given the ways in which social media has been used by extremist groups to spread disinformation, Sunstein was onto something. But two decades after he issued his warning, what does the evidence say?

Answering this question is difficult. The philosopher Joshua Cohen likened it to "asking an epidemiologist about the impact of a medical intervention on diseases of unknown etiology in a migrant population with no medical records living in an environment with a high rate of viral mutation." The task for social scientists is daunting, a fact made all the more resonant after a year of watching public health researchers struggle to make sense of COVID-19.

The good news is that this area of scholarly research has exploded in the past few years. But the evidence so far is decidedly mixed—neither a full endorsement of Sunstein's concerns nor an exoneration of the platforms and the speech they host and distribute.

Let's begin with the major worry of the pessimists: that online interactions diminish people's exposure to diverse views, encasing users in filter bubbles of the like-minded and thereby exacerbating polarization. If this view were correct, the consequences for democracy would be devastating. However, the evidence points in another direction. As one recent analysis concluded, "Even if most political exchanges on social media take place among people with similar ideas, cross-cutting interactions are more frequent than commonly believed, exposure to diverse news is higher than through other types of media, and ranking algorithms do not have a large impact on the ideological balance of news consumption." This is because social media seems to expand people's interactions with individuals outside their closest social circles—coworkers, relatives living in other places, acquaintances—who often share different, and more ideologically diverse, information.

This evidence sits uncomfortably with the popular perception that social media produces nothing but echo chambers that polarize the population. To reconcile these facts, it is important to recognize that a lot of what happens on social media involves a small minority of highly active and visibly partisan individuals. For example, a small number of very frequent users were responsible for the overwhelming majority of hyperpartisan content on Twitter ahead of the 2012 election. Another study demonstrated that a majority of people rely primarily on centrist websites for their news, and the minority who visit the most partisan websites are active news consumers and visit many other kinds of sites as well.

Even though what someone reads is shaped in powerful ways by their social network, most people's networks are actually surprisingly diverse. One group examined the ideological content of the Facebook news feeds of more than 10 million users. Though most friendship links connect people who are ideologically similar, many friends (20 percent for conservatives and 18 percent for liberals) are from different ideological groups. And approximately 30 percent of news stories in an individual's feed are ideologically cross-cutting. If you carefully compare what happens online to other modes of communication, the conclusion seems inescapable: most people are exposed to more political disagreement online than in their real-world social connections.

But how does this exposure to politically diverse views relate to polarization? A group of economists, including our Stanford colleague Matthew Gentzkow, tackled this question in two revealing ways. First, for every age cohort in the United States since 1996, polarization was measured in eight different ways, ranging from the extent of straight-ticket voting to how people reported feeling about members of the other political party. The striking finding was that polarization has increased most among the groups that are *least likely* to use the internet and social media—people in the oldest age cohorts. Based on this evidence, it's hard to conclude that social media itself is the major driver of polarization.

A more recent study tackled head-on the challenge of identifying a causal relationship between social media usage and polarization. Gentzkow and his coauthors came up with a creative plan: they offered Facebook users a cash incentive to deactivate their accounts for four weeks. This approach enabled the research team to examine whether a reduction in access to Facebook changed anything about people's informational diets and political attitudes. Because the cash incentive was offered at random to people in a large sample, they could use the existence of a cash incentive for some and not for others to figure out the effects of tuning out of Facebook. They found that most forms of polarization were unaffected by a reduction in social media usage, though people not on Facebook did have demonstrably less knowledge about current events.

Even if many of our worst fears about social media and polarization are not borne out by the data, the reality is that social media is flush with misinformation and provides an all-too-welcoming home for extremism and hateful rhetoric. And though polarization in our society might be affected by lots of factors beyond social media, the consequences of misinformation and harmful speech can be significant.

One major source of pollution in the information ecosystem is misinformation, "claims that contradict or distort common understandings of verifiable facts." Misinformation can spread in a wide variety of ways: messages from political leaders; published articles, blog posts, and tweets; even advertising. Disinformation is the subset of misinformation that is deliberately propagated to deceive. The goal is to undermine the ability of people to make a choice or judgment with clarity about the verifiable facts.

Concerns about the spread of disinformation on social media burst into public view with Russian interference in the 2016 presidential election. Teenagers in the Macedonian town of Veles were paid up to $8,000 per month each to operate approximately a hundred pro-Trump fake news sites with innocuous-sounding names such as USADailyPolitics.com. Russia's Internet Research Agency, one of the "troll factories," served as an industrial-scale manufacturer of disinformation. According to Twitter, the IRA operated 3,814 accounts, and individual operators were responsible for creating a high volume of content daily. According to one report, workers had to deliver up to fifty comments daily on new articles, up to six Facebook pages with multiple daily posts, or ten Twitter accounts with at least fifty daily tweets.

Did the disinformation they produced actually register with the sites' users? One team tracked more than 2,500 adults during the five weeks before the 2016 election. Examining their social media consumption and mapping stories to existing lists of "fake news," they estimated that more than a quarter of Americans had been exposed to at least one fake news article in that period. Though this is a large number, the actual volume of engagement with fake news sites is actually quite low. As a share of visits to "hard news" sites, less than 2 percent of visits involved engagement with a fake news article.

More troubling is that some people are much more at risk of encountering and sharing disinformation online. The most significant predictor of fake news sharing behavior is age: older people are much more likely to pass along fake news stories to their friends. Russia's IRA was particularly focused on exploiting this vulnerability; its accounts frequently impersonated hyperpartisan US citizens and shared disinformation only about 20 percent of the time. Most of the content it pushed was real material from local news sources, which was useful for building trust in the account and highlighting the particular interests and values of the partisan identity it was impersonating.

Similar to the challenge of objectionable content, how best to deal with misinformation and disinformation is not always clear. Perhaps more accurate information should be shown when misinformation is detected. But providing people with correct information doesn't always change what they think. And in some cases, it may even backfire. One series of studies showed that when individuals were provided with factual corrections that contradicted their political beliefs, they responded by becoming even more committed to their previous misconceptions. Psychologists attribute this disturbing finding to something called *motivated reasoning*, the idea that people interpret new facts about the world in ways that align with their ultimate goals. And though researchers continue to debate just how prevalent backfire effects are, there is evidence that efforts to correct misperceptions may be insufficient to override the outdated misinformation when it comes to people's judgments.

The danger here is real: when individuals hold misinformed beliefs, it can have an impact on large-scale policy outcomes. The myth of "death panels" is just one example of the way oft-repeated misinformation impacted the health care reform debate in 2009. As one expert summed it up, though an uninformed citizenry is surely not ideal, a misinformed citizenry can be quite dangerous.

As with misinformation and disinformation, hate speech has found a welcoming environment on mainstream social media platforms. Yet hate speech makes up only a tiny fraction of the content that is posted online. In one study, a research team analyzed more than 750 million tweets by political figures and 400 million tweets by a random sample of Americans between 2015 and 2017. On any given day, between 0.001 and 0.003 percent of tweets contained any hate speech—an exceedingly small portion of the content generated by US Twitter users. Of course, this likely understates the actual amount of hate speech posted, since the companies proactively take down a lot of such content before it is seen by anyone. Facebook claims to take down 89 percent of posts violating its hate speech policies before a user reports a concern. But just because a small amount of hate speech ends up being posted doesn't mean that it isn't seen. In fact, a cross-national survey of internet users aged fifteen to thirty found that more than half of Americans reported having seen hate speech online. Exposure was also high in a number of other countries, including Finland (48 percent), the United Kingdom (39 percent), and Germany (31 percent).

Those who are active propagators of hate speech are a small but energetic minority. They tend to tweet more frequently, follow more people each day, and are embedded in networks of other producers of hate speech who tend to retweet one another's content. Strikingly, they don't necessarily start out producing misogynistic and racist hate speech. In fact, the evidence suggests that they start out with softer and more indirect posts, graduating to the use of more virulent language only as they enter more extreme social networks and lose their concern about social stigma. These networks are so concentrated that one research team showed that it could accurately predict the users who were going to post anti-Muslim tweets in France after the 2015 terrorist attacks in Paris, even if they had never mentioned Muslims or Islam in their previous tweets!

This speaks to a bigger concern: for online speech to have real offline effects, it doesn't need to dominate the social media landscape. Even a small number of posts can have massive consequences for the well-being of people living in the real world, especially if they energize, incite, or mobilize a small group of individuals who want to do harm.

Dylann Roof, the white supremacist who murdered nine people at the Emanuel African Methodist Episcopal Church in Charleston, South Carolina, in 2015 drew his inspiration from online content shared by a radical Christian group. The perpetrator of the attack on the Pittsburgh Tree of Life synagogue in 2018 had been exposed to hateful content on Gab. And the killer in the Christchurch, New Zealand, mosque shootings in 2019 had reportedly been radicalized online. In these cases, we can't know for sure whether the perpetrators' exposure to online content changed their views or was in any way causally linked to their actions. But there isn't any debate about the fact that social media created a space in which they could share extremist speech and receive positive feedback.

Research is beginning to show that online hate can be a cause of real-world violence. One study in Germany showed that anti-refugee hate crimes increased disproportionately in areas with high Facebook usage when anti-refugee sentiment was prominent online. That effect disappeared when particular areas experienced internet outages or service disruptions, suggesting a causal connection between the two. Likewise, in the United States, higher Twitter usage has been associated with an increase in anti-Muslim hate crimes since the start of Trump's campaign in 2016, especially in areas that were early adopters of Twitter. Though the overall prevalence of hate speech may be rare, its concentration among particular groups—and the severe consequences of it—underscore just what's at stake in addressing the pollution of our information ecosystem.

It's a terrible predicament. We have good reasons to value free speech, democracy, and dignity, but we can't simultaneously maximize the fulfillment of all three. There are painful trade-offs to be faced in the world that technologists have built. In the wake of the 2016 presidential election and the revelations about Cambridge Analytica, technologists told us that they see the problems and recognize the tensions. They have offered apologies and assured us that they are hard at work on a fix. Yet the 2020 election was a mess as well, with its own problems: deliberate falsehoods about candidates, doctored videos, and widespread misinformation about election fraud. Some of the misinformation was tagged by tech companies, certainly improving over their performance in 2016, but much of it was not. And the events at the US Capitol two months after the election were a wake-up call about how real the dangers of misinformation and hate speech are, not only for individuals but for the integrity of our democratic institutions. Are there technical solutions to the problems of democracy-undermining and dignity-violating harmful content?

CAN AI MODERATE CONTENT?

If powerful AI algorithms can enable tech platforms to up-rank content for us to see, might it be possible to use similar technology to identify harmful content and down-rank or delete it entirely? In fact, Google, YouTube, Facebook, and Twitter (among others) do exactly that. With billions of new text, image, and video uploads each day, it's a necessity. It's even been argued that algorithmic curation of this ocean of content is the core function of a platform. For most large platforms, automated review provides the first line of defense in content moderation, attempting to identify objectionable material—nudity, child exploitation, terrorism, hate speech, bullying—that is then reviewed by human moderators. Mark Zuckerberg wrote that "Artificial intelligence . . . already generates about one-third of all reports to the [Facebook content moderation] team that reviews content for our community."

The internals of these systems, often based on the same forms of deep learning we discussed in earlier chapters, are generally not published so that system weaknesses cannot be exploited. But in the name of greater transparency, Facebook provides a Community Standards Enforcement Report that provides statistics on the results of using its internal systems—including both AI and human moderators—to find content that violates its standards. The report is framed as a race: What percentage of objectionable content did the systems find before users reported it? Organized by category, the report shows that violations of some standards are easier to detect than others. According to the February 2021 report, Facebook found 99 percent of content violating its "Adult Nudity and Sexual Activity" standards before users report it. For "Suicide and Self-Injury," that number dropped to 92 percent. And for "Bullying and Harassment," it was only about 50 percent. This variability in performance reflects the subjectivity and nuance of the material being assessed. Detecting nudity is often just a matter of finding enough naked skin pixels in an image accompanied by words that are overtly sexual.

Identifying hate speech is harder. Deciding whether to take action on a statement such as "Men are scum" requires contextual and cultural judgment. Despite Facebook's actions in that particular case, it's not just an analysis of whether certain words appear in a post but rather what those terms are supposed to mean, who said them, and who they may be directed toward. Making such determinations generally requires human understanding and poses a significant challenge to purely automated systems. In some contexts, describing a person as a cockroach is a reference or allusion to Kafka's short story "The Metamorphosis." In Rwanda, it is a dehumanizing insult and a reference to the horrors of the 1994 genocide. The situation with bullying becomes even more difficult, especially since content standards vary even more depending on the context and the individual involved. This is a point that Facebook readily concedes, stating "because bullying and harassment are highly personal by nature, in many instances, we need a person to report this behavior to us before we can identify or remove it."

Though AI will continue to improve in the future, it's really only the first line of defense. The number of human content moderators needed to tackle the volume of information on these platforms is enormous. In 2017, YouTube CEO Susan Wojcicki announced that the company would hire 10,000 content moderators in the coming year. The following year, Mark Zuckerberg wrote that "the team responsible for enforcing these [community standards] policies is made up of around 30,000 people. . . . they review more than two million pieces of content every day." Indeed, the big tech platforms should be credited for taking an aggressive stance toward content moderation and hiring armies of content moderators.

But being a content moderator comes with a price. Selena Scola, a onetime content moderator at Facebook, sued the company in 2018, claiming that the stress of the job had caused her to develop post-traumatic stress disorder (PTSD). Her suit reported that her job involved viewing "distressing videos and photographs of rapes, suicides, beheadings and other killings." The suit was joined by several other former content moderators who had had similar experiences, claiming that "Facebook had failed to provide them with a safe workspace." In May 2020, Facebook agreed to settle the case with its former and current content moderators to the tune of $52 million. Each content moderator is to receive $1,000, with additional funds going to those diagnosed with mental health disorders.

Even assuming that most objectionable content can be identified quickly, the most formidable challenge—a contest between competing values—arrives when we ask how that content should be handled. Nathaniel Persily, our colleague at Stanford, has offered a taxonomy of "Ds" that provides a wide array of options: deletion, demotion, disclosure, delay, dilution, diversion, deterrence, and digital literacy.

Deletion may seem like an attractive approach for any content that violates platform rules, but doing so allows the people posting it to realize that their content was banned and try again with different (or altered) text, images, or video. Much like search engine optimization, it can become an arms race between the platforms trying to eliminate objectionable content and malicious users—sometimes using bots—trying to promote it. Zuckerberg discussed exactly that phenomenon, writing "While we've made steady progress [in content moderation], we face sophisticated, well-funded adversaries. They won't give up, and they will keep evolving." The rise of deep fakes—fake images and videos created by complex machine learning and image processing techniques—is just one example. Though AI might be used to filter misinformation and objectionable content, it can also be used to create new forms of it, providing even greater challenges for algorithms and humans alike.

Demotion and delay are other common approaches and are more difficult to detect by the producers of the content. Objectionable content might simply be down-ranked; it may still show up in your feed, but you'll need to scroll quite a bit to get to it. The consequences are that "posts that are rated as false are demoted and lose on average 80% of their future views." AI plays an especially important role here, since down-ranking content quickly, before a human has time to review it, or choosing to delay its release can help limit its dissemination before a final human judgment can be brought to bear on its removal. That might make a critical difference in the days before an election.

Disclosure and diversion present other possibilities, namely identifying the source of the objectionable content or providing alternative information along with the original posting to help users make more informed judgments. Of course, this assumes that content tagged in this way will actually change users' evaluation of it and not have the unintended effect of highlighting inappropriate content for curious users to read. Even with the forward march of AI, it's unlikely that a technical solution alone will be enough, at least in the foreseeable future. The amount of information that's needed to make nuanced decisions about content—the writer's context, the implications the viewer may draw from an image, ongoing world events—means that machine learning models won't have sufficient input to be able to make accurate determinations about appropriateness. Many such decisions can even be difficult for humans to make.

The inability of AI alone to solve the problem was on full display at the outset of the COVID-19 pandemic. YouTube, Twitter, and Facebook wanted to limit the number of workers, including content moderators, coming into the office. Concerns for user privacy posed obstacles to accessing data from home computers and networks. Company guidelines often require that content moderators do the job only in secure locations in corporate offices. Twitter acknowledged the impact this greater reliance on AI for content moderation would have on its platform, writing in a March 2020 post that it was "increasing our use of machine learning and automation to take a wide range of actions on potentially abusive and manipulative content. We want to be clear: while we work to ensure our systems are consistent, they can sometimes lack the context that our teams bring, and this may result in us making mistakes." YouTube made a similar proclamation, stating that "we will temporarily start relying more on technology to help with some of the work normally done by reviewers. This means automated systems will start removing some content without human review. . . . As we do this, users and creators may see increased video removals, including some videos that may not violate policies."

Most users of YouTube would never know that a video they tried to upload had been erroneously removed by AI. The situation on Facebook was much more visible, as users could easily see when their posts—often with reasonable content related to COVID-19—were filtered with the message "Your post goes against our Community Standards on spam." Users erupted, some accusing Facebook of partisan censorship and others wondering what in their post could have possibly triggered its removal. Ultimately, an answer came from Guy Rosen, the vice president of integrity at Facebook, who wrote, "This was an issue with an automated system that removes links to abusive websites, but incorrectly removed a lot of other posts too." The admission, though certainly not indicative of malicious intent, shows the shortcomings of relying entirely on technology to solve content moderation problems.

A SUPREME COURT FOR FACEBOOK?

The mission statements of the big tech companies are grandiose visions of making the world a better place. But as the problems of polarization, misinformation, and hate speech have become clearer, all of the companies have acknowledged the potential risks involved in serving as platforms for online content, even if their mission statements have not changed.

As a result, the companies have continued to evolve their internal policies about content moderation. These "community standards" aim to define permissible and impermissible user-generated content and behavior, and they outline the array of actions that might be taken against users who post content that runs afoul of the standards. Aside from content that breaks actual laws, such as intellectual property violations and coordination of criminal activity, content moderation in most cases is not, strictly speaking, required by existing US legal doctrines. Most pornography is not against the law, nor is hate speech, misinformation, or disinformation. In fact, the First Amendment and the 1996 Communications Decency Act provide substantial autonomy and legal immunity to private tech platforms that host content generated by users. But the combination of the economic incentive to grow a massive user base and the recognition of the corporate responsibility to steward their platforms with an eye toward their broader social effects has led every company to dedicate significant resources to policing content. As the legal scholar Kate Klonick argued, these companies have become the new private governors of speech. And with user communities numbering in the hundreds of millions or billions of people, the small number of people setting and enforcing "community standards" are among the most powerful people in the world. As a result, in recent years, pressure has mounted for the largest companies to be more accountable and transparent as a way to earn more legitimacy for the power they wield.

In 2018, Facebook announced a new initiative to create a "Supreme Court," an independent oversight board that could review takedown decisions the company made based on its community standards. "I've increasingly come to believe," wrote Zuckerberg, "that Facebook should not make so many important decisions about free expression and safety on our own." In 2020, the company named the first forty members of the Oversight Board, a roster of scholars, civil society leaders, and former elected officials from across the globe. The operation of the board is funded by an independent trust that Facebook established, creating another layer of independence from the company. Like the actual US Supreme Court, the Oversight Board has the power to decide whether to accept an appeal. According to its current charter, the decision of the independent body is binding on Facebook. So if a user brings an appeal to the board and wins the case, Facebook would be required to restore the content.

The board began meeting in late 2020 and announced its first decisions in early 2021. In four of five cases, the board overturned Facebook's removal of content, signaling a will to challenge the platform and establish the authority of its external review. Then it agreed to hear a case with global repercussions: whether Facebook had been justified in indefinitely suspending Donald Trump's account in the wake of the insurrection at the US Capitol on January 6, 2021. In late spring 2021, it upheld Facebook's decision to suspend Trump but rejected an indefinite ban. It gave Facebook six months to revisit the case and provide clear, public standards for any continuing ban. A Solomonic decision, it pleased no one and returned power to Facebook. An odd choice if the Oversight Board was intended to diminish the unchecked power of Facebook in deciding the boundaries of permissible speech on its platform.

It remains to be seen whether the Facebook "Supreme Court" will succeed, but it represents an important first step in creating at least one new mechanism of accountability and transparency that could serve to stimulate wider public conversation about the governance of private platforms. At the end of the day, the Oversight Board is still an effort at self-regulation. It has operational independence, but the initial roster of judges was selected by Facebook, the charter of the board was created by Facebook, and the decisions of the board apply only to Facebook, not to other companies. What's more, the board's role is designed to be quite limited; the job is not a full-time position for its members, so they have the capacity to hear only a very tiny fraction of total appeals. And the board's decisions concern only actions regarding content, not the broader principles of the community standards policy. Skeptics might believe that it is a diversionary tactic to draw attention away from Facebook's other problems with antitrust laws and privacy. Or that its creation is nothing more than a convenient way to blame the Oversight Board for future complaints about content that is removed. But as its initial decisions reversing Facebook reflect, the board is establishing its independence. Perhaps over time it will acquire greater legitimacy, serving, as its creators at Facebook hope, as a model for other companies. In the words of Nick Clegg, a former deputy prime minister of the United Kingdom and current vice president for global affairs and communications at Facebook, it may "even be co-opted in some shape or form by governments."

MOVING BEYOND SELF-REGULATION

It's easy to understand why Mark Zuckerberg and Jack Dorsey don't want to be the arbiters of speech on social media. The challenge they faced in 2020 as Twitter and Facebook grappled with regulating content about COVID-19 and the presidential election revealed the difficulty of the judgments they need to make. But does it make sense for government to get involved in the regulation of speech?

Most Americans would shudder at the thought. Indeed, the First Amendment of the Constitution couldn't be clearer: "Congress shall make no law . . . abridging the freedom of speech." For all of their limits, some platforms are making a sincere effort to manage content online—whether because of their particular values, their economic incentives to maintain a satisfied base of users and advertisers, or their sense of corporate responsibility. They are even thinking hard about the legitimacy of their decisions on content, as Facebook's experiment in independent oversight suggests. Yet this approach to managing the effects of online content has an Achilles' heel: it relies on the actions of a decentralized set of profit-driven firms to protect something in which we all share a profound interest: the health of our democracy. It does not make sense for the future of a healthy public sphere to rest entirely in the hands of a small number of powerful companies. Governments must get involved—and in so doing balance the importance of free expression against other essential goals.

The first issue is whether there are particular forms of speech that government should prohibit. Most democratic governments have shied away from significant restrictions on speech because free expression is so essential to a thriving democracy. Indeed, those responsible for the growth of democracy globally were often political and social revolutionaries who challenged—at great risk to themselves—the speech restrictions imposed by monarchs, dictators, clerics, and military rulers.

In the United States, the answer to "dangerous speech" has always been more speech. The concern about protecting the right to speak from the intrusion of government has almost always taken precedence over the potential harms that speech might bring about. The Supreme Court has called out a narrow set of exceptions, such as cases where speech is "directed to inciting or producing imminent lawless action and is likely to incite or produce such action." But in practice, the Court has severely limited even these exceptions. In perhaps the most famous case on this issue, the Court reversed the conviction of a Ku Klux Klan member found guilty of advocating violence to accomplish political reform at a rally in Ohio in 1964. At a public event with armed Klansmen, the Klan member had said, "If our President, our Congress, our Supreme Court, continues to suppress the white, Caucasian race, it's possible that there might have to be some revengeance [sic] taken," but the Court was not convinced. The justices concluded that the First Amendment requires a distinction between advocating a point of view and inciting immediate violent action. Other Supreme Court decisions have made clear that US law protects speech motivated by an intent to attack or disparage a person based on his or her race, gender, and religion. As a result, virtually no speech in the United States is illegal, unless that speech is linked convincingly to the incitement of imminent violence.

The United States is an outlier in its absolutist commitment to free speech, even among the most enduring democracies. Germany is a famous example of an alternative approach. Following the horrors wrought by Nazism and widespread anti-Semitism, the German government began to build a strong social consensus around the censorship of certain kinds of speech. Racist and anti-Semitic rhetoric were at the top of the list, but so, too, were Holocaust denial, antiforeigner sentiment, and even insults and blasphemy. Hateful speech could be punished with a prison sentence, and calling a politician a liar might invite a defamation suit. Germans have grown accustomed to these constraints and largely accepted them as part of what they have come to call a *wehrhafte Demokratie*, or "democracy capable of defending itself."

Germany is not alone. In Canada, the overriding concern of hate speech laws is to prevent discrimination. So although freedom of expression is cherished, Canada recognizes that there are times when the right to speak freely should be constrained because it undermines people's right to be treated equally. This means that Canada's laws have a much more forgiving standard for censoring speech. It is sufficient to show that comments incite hatred or discrimination against identifiable groups, even if they do not result in violence or threats to public order. Though enforcement varies, other countries, including the United Kingdom, Ireland, Brazil, and India, as well as the European Union, have restricted incitements to hatred and discrimination even if there is no evidence that violence might result.

Though the social media platforms must already stay attuned to these differences in law, they are paying increasing attention in the wake of a new German law related to online speech. Passed in 2018, the law requires companies to take down "manifestly" illegal content, including hate speech, within twenty-four hours or to be subject to significant corporate and *personal* fines. That change forced Facebook and other companies operating in Germany to double down on their content moderation capabilities lest they have to face a barrage of government actions. Other countries, including Austria and the United Kingdom, have begun to debate similar measures.

Whatever your view about the permissibility of hate speech, it's worth noting that, as our colleague Renee DiResta at the Stanford Internet Observatory has written, a commitment to freedom of speech does not guarantee anyone freedom of reach. Though a commitment to freedom of speech allows individuals to speak their mind, there is no right to algorithmic amplification. Just as no one has a right to have his or her crackpot opinions published in a newspaper, neither is anyone entitled to have his or her posts retweeted, amplified, or recommended. This is what makes text messaging fundamentally different from posting content on a social media platform. We have strong expectations that free speech should permit us to communicate with others directly and without censorship via messaging services. But invoking free speech does not mean that the government or any company needs to provide you with a megaphone, whether literal or algorithmic.

But if companies have wide latitude to limit the algorithmic amplification of potentially harmful speech, when, if ever, should democratic governments require bans or limits on online speech? In weighing the value of restrictions, we are guided by a view of democracy whose goal is not to generate the "best" or "correct" outcomes but to establish guardrails that ward off the worst. The implication is clear: government restrictions on speech should remain rare and circumscribed. Calls to legally curtail forms of expression that perpetuate stereotypes, use socially unacceptable labels, or challenge group status must not be taken as easy solutions, even in the age of the internet. This may be uncomfortable to accept, especially at a time when concerns about racial bias, discrimination, and systemic racism dominate public debate. But the risk that government censorship will be misused can create even greater problems for democracy. In places where hate speech laws have been implemented, including Germany, the best evidence suggests that restrictions merely constrain what is discussed in the most public places, driving underground the noxious hate speech and disparagement that the laws are designed to erase.

One important idea for how to rethink free speech in a digital age is actually not about imposing greater censorship but rather for the government to ensure that the right to speak freely is actually available on an equal basis to all. According to Tim Wu, the role of the government and the law is "defending the principal channels of online speech from obstruction and attack, whether by fraud, deception, or the harassment of speakers." This requires that law enforcement take responsibility for preventing, deterring, and sanctioning private actors who attempt to silence speakers.

We know what these organized efforts look like: online trolling designed to humiliate, harass, and discourage targeted speakers; defamation campaigns that use fabricated stories and rumors to damage a speaker's reputation; swarmlike attacks over email, telephone, or social media to punish particular speakers. Some of these tactics draw significant public attention, such as the accusation that Hillary Clinton was involved in a pedophile ring run out of a Washington DC pizzeria. But most of the trolling happens in private, where the victims experience significant harm that goes largely unnoticed.

According to a 2021 Pew survey, four of every ten Americans and a majority of young adults have experienced online harassment. The experience falls disproportionately on minority and marginalized groups, with a large majority of gays, lesbians, and bisexuals reporting that they have been targeted with online abuse. There is also a long history of online sexual harassment and abuse of women, and it can be shockingly graphic. Laura Bates founded the website Everyday Sexism Project, where women can share their own experiences with sexism. After creating the site, she was targeted online, receiving over two hundred abusive messages a day. "The psychological impact of reading through someone's really graphic thoughts about raping and murdering you is not necessarily acknowledged," she told Amnesty International. "You could be sitting at home in your living room, outside of working hours, and suddenly someone is able to send you an incredibly graphic rape threat right into the palm of your hand." The frequent result of such online abuse is to drive people offline, undermining equal access to the communicative tools of the internet and rights to free expression to which all are entitled.

We need to push the government and platforms to work together to protect free speech rights from such threats. Though some responsibility for this rests with the platforms, government has a critical role to play. Its task is not to guarantee high-quality public debate but to protect the channels of public debate from deliberate attack. This begins with robust enforcement of existing federal and state laws that prohibit harassment and cyberstalking and laws governing fraud, deception, and identity theft that can be used to fight deceptive propaganda campaigns. But new regulations might also be required, including dedicated anti-trolling laws to police mob-style attacks on journalists, public figures, and ordinary citizens. In this case, the government's role is not only about flagging harmful content but about taking enforcement actions against those who produce it. Though American free speech purists might resist moves in this direction, they are what's required to preserve the right to free expression in the digital age.

THE FUTURE OF PLATFORM IMMUNITY

Addressing the pollution of our information ecosystem also means figuring out what obligations companies should have beyond fealty to their shareholders and adherence to existing laws. In short, should government hold companies to a higher standard of behavior in order to protect democracy?

Any conversation about corporate behavior in the United States must begin with Section 230 of the Communications Decency Act, known as CDA 230 for short. This provision is nothing short of the oxygen that has enabled internet platforms to grow. Passed in 1996, CDA 230 immunizes providers of interactive computer services for liability arising from user-generated content. More specifically, it says, "No provider or user of an interactive computer service shall be treated as the publisher or speaker," thus enabling platforms to facilitate the posting and sharing of content without significant concern about legal liability. The law is as generous toward the providers of computer services as it could possibly be. It protects companies from lawsuits if they leave offending content up and protects them from litigation if they take content down—the so-called good Samaritan provision.

Here's an example: If a YouTube user were to post a defamatory video, the user could be sued but YouTube could not. Without CDA 230, websites such as Facebook, Twitter, Instagram, and YouTube would have to exercise strong judgment over everything users produce, as if they were newspaper editors—something that is technically difficult these days. In the case of smaller social networking platforms such as Parler, which tout the promotion of free speech and lack both the will and the substantial resources needed to provide large-scale content moderation, CDA 230 serves an even more critical role in providing them legal protection.

CDA 230 had its origin in the response to a court case decided one year earlier, *Stratton Oakmont, Inc. v. Prodigy Services Co.* In this case, Prodigy, an early provider of internet access and information, was found "liable as a publisher for all posts made on its site *because it actively deleted some forum postings*" (authors' emphasis). The court considered Prodigy to be more than a simple distributor of content because its use of automated curation tools and guidelines for posting were a "conscious choice, to gain the benefits of editorial control." The decision sent shock waves through the growing internet business, raising fears that the new platforms would become targets of lawsuits. In response, a bipartisan effort in Congress added what became Section 230 to a bill already under consideration to regulate the access of minors to indecent online content.

Daphne Keller, a former associate general counsel for Google, is one of the world's leading experts on the issue of "intermediary liability," the technical term for the extent to which platforms bear any legal responsibility for the content that they post, share, or curate. When considering something such as CDA 230, she says, we have to recognize that we are balancing three goals: (1) the prevention of harm, (2) the protection of lawful speech and online activity, and (3) the prospects for innovation. In the 1990s, legislators were looking for ways to enable content moderation without running into the legal issues that had gotten Prodigy into trouble. Ultimately, the law created incentives for platforms to leave as much content up as possible, yet also protected them in the event they decided to take something down. But with the rise of misinformation and disinformation, as well as hateful and extremist rhetoric, is this a healthy balance of these values today?

A lot of people are calling for a fundamental rethink of CDA 230. During his presidential campaign, Joe Biden called for revoking CDA 230 immediately in an interview with the *New York Times* editorial board. Referring to Facebook, he said, "It is not merely an internet company. It is propagating falsehoods they know to be false. . . . There is no editorial impact at all on Facebook. . . . It's totally irresponsible." Though his frustration is understandable, the danger of this path is that it threatens free expression by changing the immunity framework in a way that would give companies an incentive to take down any content that might be thought offensive and could trigger a lawsuit. Such a move could also crush small and medium-size companies, which have far fewer resources than today's tech giants, with the weight of responsibility for content removal, thus undermining innovation.

The critics come from the right as well. One of the most prominent enthusiasts for reform is Josh Hawley, a forty-year-old Republican senator from Missouri. Ironically, Hawley is probably best known as the first senator to object to certifying the results of the 2020 presidential election and for raising his fist in support of pro-Trump protestors before they stormed the Capitol. As a new senator, he launched a crusade to rein in the power of the platform companies. His particular concern is the censorship of conservative voices and right-leaning content. If there is going to be content moderation, he wants to ensure that moderation is "politically neutral"—and he's prepared to hold platforms liable if their takedowns of content are biased. One piece of legislation he authored gives individuals who believe they are being unfairly censored the right to sue companies for at least $5,000 as well as attorney's fees. One can imagine the flood of lawsuits this legislation would unleash.

In this partisan back-and-forth, what's really at stake is whether we should expect internet companies to perform the same kinds of editorial functions undertaken by traditional media organizations, like newspapers, radio, and television. There are definitely reasons we should consider tech platforms publishers of content. Unlike the telephone company, which simply hosts a connection between two speakers, the companies are engaged in active curation of what users see and hear through algorithms tuned to maximize engagement. And they are dominant providers of communication channels in their space, with little competition. This looks and feels like the broadcast networks of the last century before the growth of cable television, with the critical distinction that unimaginably more people are able to broadcast.

Two additional things are different now, both of which matter for the prospects of building an effective regulatory regime for internet platforms. The first is the level of partisan polarization. It's difficult to imagine parties on opposite ends of the political spectrum coming to agreement on what constitutes fake news and misinformation, when hateful language becomes incitement to violence, and whether a curation process is politically neutral. The second is that legislators historically justified the regulation of broadcasting in terms of the physical limits of airspace that created spectrum scarcity. Because there wasn't enough broadcast spectrum to enable a fully competitive market, broadcasters were expected to serve the public interest as well as their own commercial interests. That led to television practices that many of us took for granted, such as balanced coverage of political issues and content relevant to local communities. Though internet companies dominate the new channels of communication, their market dominance isn't a function of scarcity, and it could in principle be challenged by other companies.

If we can't collectively agree on what content should and should not be permitted online, it's important that we have a diverse and competitive marketplace of internet platforms. Seeking to remain competitive in attracting and retaining users, internet platforms will continue to moderate content themselves. If the platforms become overwhelmed by hate speech and fake content, some users may seek out alternatives—as will some advertisers, the platforms' key revenue source. This dynamic already played out once in the early days of the internet as pioneering search engines such as AltaVista, Lycos, and Excite became clogged with spam web pages more intent on selling products than on providing users with the information they were looking for. Those early search engines were quickly displaced by Google, whose new search technology was much more effective at down-ranking or eliminating spam pages in the results users received. It's not only users who have agency; advertisers can also exercise influence on platform behavior. For example, a number of companies agreed to pull ads from Facebook in 2020 as part of the #StopProfitForHate campaign when the company refused to take down or limit exposure to Trump's post "When the looting starts, the shooting starts." Of course, the inverse is also possible. Some users might choose platforms with looser content moderation, as we saw with the migration to Parler after the deplatforming of President Trump. However, the data so far suggest this will be an attractive option for a relatively small share of total users.

Government should get involved when organized efforts to spread misinformation and disinformation threaten the integrity of the democratic process. We already recognize an appropriate role for government when we want to stop child pornography, human trafficking, copyright violation, and radicalization. In these cases, government creates a set of rules and expectations that shape how companies carry out content regulation. The Digital Millennium Copyright Act has served as the primary source of guidance for copyright violation on the internet and provides significant safe-harbor provisions to give internet companies good reason to enforce its provisions. This brings democratic legitimacy to the process by which content is removed and ensures that we aren't totally reliant on the goodwill and good judgment of CEOs.

It's time to do the same to protect democracy. Of course, it's unrealistic to legislate for high-quality debate or a "fairness doctrine" online that would ensure equal airtime for competing views. That doesn't even exist at our Thanksgiving dinner tables. But it is realistic to think that we can pursue some commonsense reforms. The debates about the future of CDA 230 offer an opportunity to provide stronger incentives for companies to take action on disinformation and misinformation, either by legislating new limits to the immunity of platforms or by preserving blanket immunity on the condition that the companies take more assertive action to protect democracy and report transparently on their progress.

For example, it is already illegal for foreign interests to engage in election-related advertising in the United States. The problem is that platform companies have not been especially good at identifying and preventing this kind of behavior. We need strong incentives for companies to eliminate such activity from their systems, either via the threat of legal action or as a requirement for maintaining their immunity. We also need to incentivize companies to protect users from fraud, such as when paid agents purporting to be genuine users or unlabeled bots flood the information ecosystem with content designed to manipulate or mobilize. Transparency requirements could be used to ensure that platforms disclose information that is relevant to help users evaluate the credibility of sources, such as who is paying for political ads. Warnings can also be encouraged for content that approved fact-checking organizations have disputed, as we saw done in the 2020 election. Right now, these decisions are left entirely to platforms and their own internal rules. A democratic process could provide more legitimacy to contested judgments and create standards to which all platforms would adhere. And given the partisan climate, it also seems important to bring greater transparency to the platforms' content moderation policies in general with mandated (and standardized) reporting on their processes and practices.

Of course, these modest regulatory steps don't preclude more dramatic shifts on the part of the companies themselves. Platforms could adapt their curation algorithms to take into account objective indicators of "credibility" as they promote and rank information. They could restructure the presentation of information to ensure that conflicting or diverse perspectives are presented together at the top of search results or on a news feed. They could decide that particular practices that many believe are harmful to our democratic public sphere will no longer be permitted on their platform, such as the "microtargeting" of political advertisements at individuals on the basis of highly granular information about their likes and dislikes. They could decide not to sell political advertising during critical time periods or at all. They could hold public officials to the same standards as any other users on their platform and create a system of penalties for users who repeatedly pollute the information ecosystem, leading to account deactivation for the worst offenders. And they could introduce friction into the system to reduce virality, especially when they have identified organized efforts to spread misinformation and disinformation. But for any of these more fundamental changes in platform behavior to emerge organically, we would need a more competitive market in which users can choose which platforms to use based on the different moderating or curatorial services they provide.

CREATING SPACE FOR COMPETITION

How to achieve a more competitive marketplace online is the final piece of the puzzle. The extent to which the existing internet platforms dominate online communication is staggering. Google is responsible for over 90 percent of online searches worldwide. Facebook generates nearly 70 percent of all visits to social media sites on a monthly basis, with Twitter accounting for an additional 10 percent. Facebook, YouTube, WhatsApp, and Instagram all boast more than 1 billion monthly active users; they are the world's largest social media platforms. Facebook, WhatsApp, and Instagram are all part of the same company. And YouTube is part of Google.

The implication of this market dominance is that CEOs Mark Zuckerberg, Jack Dorsey, and Sundar Pichai wield an outsize influence in determining the extent to which the pollution of our information ecosystem will proceed uninterrupted. Their choices about whether and how to moderate content disproportionately impact the information we consume and the health of our digital public sphere, especially in the absence of government intervention. And given this degree of market concentration, it is far easier for propagators of fake news and misinformation to be successful; they simply need to game one or two algorithmic systems successfully in order to reach millions of people.

The promotion of a healthy, thriving digital public sphere depends on ensuring that Facebook, Twitter, YouTube, Google, and others face healthy competition from other companies that offer high-quality services with different approaches to content moderation. That isn't happening now. To make it happen, we would need to radically update the way in which antitrust enforcement works in the United States and develop a more coordinated approach with Europe.

In the United States, at least, antitrust enforcers have been totally confused about how to handle the major social media platforms. With their mandate to ensure that consumers pay a fair price for products, the enforcers are out to sea when it comes to regulating companies that give away their products for free. The Europeans have thrown the book at

Google, operating upon a much broader view of what constitutes anti-competitive practices. Successive antitrust actions have targeted Google's platform dominance by challenging the privileged placement of other Google services in Google search results, the company's requirement that advertising partners not do business with any of its competitors in search, and the way in which Google has used its Android platform for mobile phones to lock in the installation of other Google apps.

In 2021, President Biden nominated Lina Khan, one of the leading voices in the US for a reinvigorated antitrust approach, to the Federal Trade Commission (FTC), signaling that the US may be following Europe's lead. That could mean returning to the progressive origins of antitrust policy in the industrial age and the concern with not only price but also market dominance. As Senator John Sherman, the author of the first US antitrust act, put it, "If we will not endure a king as a political power, we should not endure a king over the production, transportation, and sale of any of the necessities of life." We should take a set of concrete steps that ensure the existing platforms confront healthy competition.

The first step involves a clear regulatory commitment to maintaining nondiscriminatory and equal access to the internet. In 2015, the US government took an important step in this direction when the Federal Communications Commission (FCC) established a regime of "net neutrality." In practice, this involved designating the major internet service providers (ISPs), such as Comcast and AT&T, as private companies with public obligations to ensure equal access. This means that your internet service provider can't control or manipulate how you use the internet, for example by forcing you to get your news from one source or to use a particular search engine. The FCC quickly reversed this commitment to net neutrality after Trump took office, which was a big win for a small set of powerful companies. But the battle is far from over, as the issue winds its way through the courts, Congress considers new legislation, and President Biden puts his imprint on the FTC and FCC.

Governments also need to put a stop to the platforms' leveraging their power in one market to monopolize a second. One of the major EU antitrust cases targeted exactly this issue, in which Google search results favored Google's own comparison shopping website while pushing those of its competitors down to page 4 of the search results. Though we should celebrate Google's success in developing an extraordinary search engine, its dominance of the search market does not entitle it to use that monopoly power to undermine other product and service competitors. What's needed is a "separations" regime—an approach pioneered during the Progressive Era to stop the railroads from dominating other areas of commerce simply because they owned the railroad tracks.

Finally, we need an aggressive strategy of preventing and reversing anti-competitive mergers and acquisitions. The fact that big tech platforms are so dominant—and so rich—means that they simply buy up any competitors that represent a threat to their position. An example of such a merger was the acquisition of Instagram by Facebook. Although few people saw the risk at the time, the merger allowed Facebook to lock in its dominance of social media by scooping up one of its most significant and fastest-growing competitors.

Although the politics of regulation has an important role to play in creating healthy market competition in search and social media, the reforms we discussed in chapter 5 regarding privacy rights are also important. If users gain the right to transfer their data from one platform to another, the likelihood that new companies can challenge the dominance of existing platforms goes way up. As long as companies are able to lock up the information that people upload and share in a single platform, it will be difficult for innovative challengers to gain a foothold.

Increased competition has the potential to better align what tech platforms do with what their users really want and, just as important, provide different options for people with different preferences. Consumers would have the option of choosing a social network or search engine that privileges credible information, removes harmful content, and protects them from fraud and abusive trolling. Right now, consumers have few alternatives—and even if they did, they wouldn't be able to bring their data with them. And the bad actors seeking to manipulate the information ecosystem would have to operate across multiple different platforms, rather than needing to game only one or two systems.

As we've seen over and over, it's not realistic to count on firms acting in their own self-interest to protect the things we value. If we care about protecting the integrity of information ecosystems, governments will have to play a role. Though we can respect a uniquely American approach to free speech in the process, Europe's efforts to ensure healthy competition provide a path forward—one that has its roots in the great traditions of the Progressive Era but can be modernized to deal with today's challenges.

PART III

RECODING THE FUTURE

Only if we acknowledge technology's power to shape our hearts and minds, and our collective beliefs and behaviors, will the discourses of governance shift from fatalistic determinism to the emancipation of self-determination.

—Sheila Jasanoff, *The Ethics of Invention*, 2016

CHAPTER 8

CAN DEMOCRACIES RISE TO THE CHALLENGE?

In early 2019, a young company called OpenAI made an announcement that immediately caused waves in the scientific community. OpenAI had created an extremely powerful AI-driven tool called GPT-2 (Generative Pre-trained Transformer, model 2) capable of generating surprisingly high-quality text. It does so with nothing more than a minimal prompt; a single sample sentence will do, such as "Write an essay about Toni Morrison's *Beloved*." The GPT-2 language model is extremely flexible, able to translate, answer questions, and summarize and synthesize other texts in addition to generating text of many different kinds, including remarkably plausible poetry, journalism, fiction, academic papers, essays for middle school, and even computer code.

What really surprised the AI community, however, was not the model used in GPT-2, the architecture of which was based on simply predicting the next most likely word based on all the previous words in the text. OpenAI's achievement was that it had scaled the system up to a new level by analyzing text from more than 8 million web pages. The striking thing was the announcement that OpenAI would not release the model, contrary to a trend toward transparency in the research community. "Due to our concerns about malicious applications of the technology," the OpenAI team wrote, "we are not releasing the trained model. As an experiment in responsible disclosure, we are instead releasing a much smaller model for researchers to experiment with, as well as a technical paper."

OpenAI was created in 2015 as a nonprofit organization funded by wealthy technologists, including Elon Musk, Peter Thiel, Sam Altman, and Reid Hoffman, who were concerned with charting a path toward safe artificial general intelligence. With a social rather than profit-making mission, the team worried that the powerful tool it created could easily be put to illicit or even nefarious use producing fake text analogous to deep-fake images and videos. Middle school students could ask it to write short essays, leading to widespread and undetectable cheating. At the extreme, propagandists could use it to create automated fountains of disinformation, delivered through fake websites and social media accounts. But what seemed a sober precaution was considered by some in the AI world either as running afoul of research norms and rank hypocrisy given the "open" part of OpenAI or as a cheap publicity stunt designed to call attention to the organization. Some AI scientists facetiously said that they, too, had made breakthrough discoveries in the lab but could not share details due to their concerns about bad actors.

By late 2019, OpenAI decided to release the full-scale GPT-2 model—with 1.5 billion parameters—as part of a staged release plan. OpenAI's scientists also reported results from their research partners, shedding more light on the earlier concerns. A study at Cornell found that "humans find GPT-2 outputs convincing." More worrisome was the work at the Middlebury Institute of International Studies' Center on Terrorism, Extremism, and Counterterrorism, which found that "extremist groups can use GPT-2 for misuse, specifically by fine-tuning GPT-2 models on four ideological positions: white supremacy, Marxism, jihadist Islamism, and anarchism." OpenAI's initial concerns seemed well founded indeed.

About a year later, the OpenAI team announced GPT-3, an exponentially more powerful next-generation model with more than one hundred times as many parameters as the largest model used in GPT-2. The deep learning ninety-six-layer neural network that forms the basis of GPT-3 was trained on a huge volume of text taken from the internet in addition to an enormous library of books and the entirety of Wikipedia. To give a sense of scale, the training data for GPT-3 is nearly 45 terabytes in size, or more than four times the estimated size of all the printed material in the Library of Congress in 2000.

GPT-3 represents an important frontier in AI research. The power of the model is undeniable, with some calling it the closest thing yet to artificial general intelligence. Without having been trained on any specific topic, it can generate convincing text based on an enormous variety of prompts. To give one example of its range and seeming ability to understand nuance and humor, consider this:

> **Kanye West Exclusive—Why He's Running for the Presidency, and What His Priorities Would Be as President.**
>
> **Prompt provided by a human:**
> *Anderson Cooper:* Yeezy, how would you describe your presidential platform?
>
> **GPT-3 output:**
> *Kanye:* My platform is very simple. I want to make everything awesome again.
> *Anderson Cooper:* But what about foreign policy?
> *Kanye:* Who cares about that stuff? Let's talk about important stuff, like Kim Kardashian's butt.
> *Anderson Cooper:* That's a bit sexist.
> *Kanye:* I'm sorry. I'm not sexist. I love women. I'm just an equal opportunity objectifier.

Despite the eerily plausible-sounding text produced, GPT-3 does not have an actual understanding of the text it generates. It's simply producing an output based on the large volume of data on which it was trained. In fact, many researchers doubt that machines will ever reach true human-level intelligence.

Mere, unenlightened humans might be impressed by the abilities of simple deep learning programs, but when looked at in a more holistic manner, it all adds up to . . . well, nothing. They still don't exhibit any trace of consciousness. All of the available data support the notion that humans feel and experience the world differently than computers do. While a computer can beat a human master at chess or Go or some other game of structured rules, it will never be able to truly think outside of those rules, it will never be able to come up with its own new strategies on the fly, it will never be able to feel, to react, the way a human can. Artificial intelligence programs lack consciousness and self-awareness. They will never be able to have a sense of humor. They will never be able to appreciate art, or beauty, or love. They will never feel lonely. They will never have empathy for other people, for animals, for the environment. They will never enjoy music or fall in love, or cry at the drop of a hat.

Actually, that last paragraph was not written by us. It was generated by GPT-3 in response to the prompt "Why deep learning will never truly X." It can also craft Harry Potter stories in the style of Ernest Hemingway, invent plausible conversations between famous people in history who never met, summarize movies with emojis, write poetry, and much more.

The reason we know about these capabilities is that OpenAI released the GPT-3 model to interested parties, albeit through an application process in which OpenAI controls access. Those granted access began playing with it and posting their findings. OpenAI announced its intention to offer GPT-3 as a revenue-generating commercial product in limited contexts. In the months between announcing GPT-2 and GPT-3, OpenAI found itself needing investment capital and converted from a nonprofit to a for-profit corporation. It promised to adhere to its social mission by pursuing an unusual "capped profit" model, by which investors in the company could get returns up to a specified cap and any additional profits beyond that would be reinvested into OpenAI's pursuit of safe artificial general intelligence. The company then struck a deal with Microsoft, which invested $1 billion in order to obtain an exclusive license to use the capabilities of GPT-3 in its products. OpenAI acknowledges the potential for malicious use, as well as the possible displacement of human labor by its text-generating machine. And as is true of other algorithmic models, OpenAI is concerned about issues of fairness and bias. But as yet there is no external oversight of GPT-3 and not even much public understanding of the tool. In essence, no rules except those adopted by OpenAI's own team regulate permissible uses of the model.

GPT-3 is the latest arrival in systems that can produce what researchers call "synthetic media" or deep fakes, the ability of increasingly powerful machines to generate or alter text, audio, images, and video in ways that are readily believable to humans. And far from resting in the hands of only a powerful few, many of these tools are, or soon will be, commercially available. And as the cost of computing resources continues to decrease exponentially, such systems will eventually become accessible to nearly anyone.

Synthetic media raises, in especially powerful form, some of the same questions and problems we have seen throughout this book. What becomes of our informational universe and our ability to trust our senses of hearing and sight when intelligent machines can automate production of media that seem authentic? What might become of human and social welfare if machines can replace human labor across many different occupations? How can we ensure that new and powerful technologies do not create or amplify existing bias or discrimination? How can we harness the benefits of advances on the technological frontier while minimizing or eliminating the risks?

SO WHAT CAN I DO?

In the blink of an eye, our relationship with technology changed. We once connected with family and friends on social networks. Now they're viewed as a place for disinformation and the manipulation of public health and elections. We enjoyed the convenience of online shopping and the unfettered communication that smartphones brought us. Now they're seen as a means to collect data from us, put local stores out of business, and hijack our attention. We shifted from a wide-eyed optimism about technology's liberating potential to a dystopian obsession with biased algorithms, surveillance capitalism, and job-displacing robots.

It's no surprise, then, that trust in technology companies is declining. Yet too few of us see any alternative to accepting the onward march of technology. We have simply accepted a technological future designed for us by technologists.

It need not be so. There are many actions we can take as an initial line of defense against the disruptions of big tech in our personal, professional, and civic lives. Perhaps the most important first step is one you've already taken by getting to this point in the book, which is to inform yourself about the myriad ways technology impacts your life. To fight for your rights in high-stakes decisions, you need to understand whether an algorithm is involved. In contexts such as being denied a mortgage, losing access to social services, or encountering the criminal justice system, you may have a right to seek more transparency into the decision-making process, and that can include determining if and how an algorithm was used. Indeed, a growing number of lawyers are finding success at uncovering the use of algorithmic decision-making and successfully challenging unjust results in the courts.

In the realm of data collection, your individual rights have been growing in recent years thanks to legislation such as the General Data Protection Regulation (GDPR) and the California Consumer Privacy Act (CCPA). The next time you get a pop-up on a website asking you to "Accept cookies," read the warning and be deliberate about what information you actually want to provide to the site and potential advertisers or whether you want to go there at all. You can also set your Web browser to reject all cookies, making it more difficult for websites to track information about you or build a profile of your behavior over time.

In his 2018 book, Jaron Lanier provided "ten arguments for deleting your social media accounts right now." And the popular 2020 documentary *The Social Dilemma* sensationally cast social media as a deliberately engineered form of addiction that could be used to control users' actions and emotions. Such characterizations lead us to believe that the only way for us to wrest technology's control over our lives is to disengage with it altogether. To quit cold turkey. Of course, that is always a possibility in the extreme, but such a view misses the point that there are benefits to be gained by engaging with such technologies if we reassert our control over them. On a personal level, we can make choices about what technology platforms we want to use and how we specify our settings for privacy and information sharing on those sites. We can take a more critical view of the information we see on such platforms, knowing that they are used not only by our friends and family but also by malicious actors who are seeking to spread misinformation. But ultimately, we can't—and shouldn't have to—rely on personal action alone to confront the disruptions caused by big tech. As we've argued throughout this book, we need to bring our collective action to bear if we want technology to respect the wider and richer set of values we want to uphold.

As a simple analogy, consider the decisions we make about driving. As individuals it behooves us to take precautions to drive carefully: watch out for pedestrians and other cars, maintain a safe speed, and so on. But we shouldn't expect that our personal decisions alone will be enough to ensure road safety. The reason we can expect safety while driving is that our individual judgment is coupled with enforceable rules of the road: traffic laws, speed limits, stoplights, and many other affordances that are the results of government regulations. Though at times abiding by rules may make our commute a little slower, the heightened safety it gives us is well worth the trade-off. The argument that we could ensure road safety by simply not driving just underscores the fact that disengaging with a system loses us the significant benefits that it can provide. In many ways, the future of navigating the information superhighway will parallel the way we've chosen to navigate real highways: we'll need to be personally vigilant, but we should also call for broader government action to create a system that puts our collective values at the fore.

IT'S NOT JUST YOU, IT'S US

The simple truth is that when problems are systemic, the solutions cannot rely only on individual action. The combination of the technologists' optimization mindset, the aspiration to maximize profit and scale, and the market dominance of a few companies defines the core problem. We have allowed the visions of technologists and their revolutionary innovations to disrupt not just marketplaces but the values that we hold dear and are core to the healthy functioning of democratic societies. What we confront today in big tech's disruptions is not a question about pop-up boxes concerning our privacy choices or whether to delete Facebook. Systemic problems require system-wide solutions. And such solutions are the traditional province of government, not consumer response; collective, not individual action.

"Individual action is great, but count me as skeptical that we will collect enough of us to change the behavior of the biggest and most powerful companies in the world." Those are the words of US Senator Brian Schatz, one of the growing number of politicians across both parties who is looking to craft policies to regulate technology companies. And he's right. Though tech companies might prefer that we navigate the choices they present to us on our own, it would be far better for us to organize our collective power to get the outcomes we want.

Technological innovation moves quickly, and the pace of change is only accelerating. Most of us can't be expected to understand the details of emerging technologies or become experts on AI. So the real problem we face is not to stay abreast of the latest technological developments but rather to determine how to balance the competing values that come into play as innovation renders new possibilities and choices.

It's time that citizens engage in a vigorous debate about the values we want technology to promote, as opposed to settling for the values that technology and the small group of people who produce it impose on us. Doing that will require our democratic and civic institutions to work together with tech companies to infuse a broader set of values into how technology is developed and deployed.

The COVID-19 pandemic revealed just how many digital tools and services, such as videoconferencing, have become essential to our lives. And there was certainly civic-minded engagement by tech companies, as search engines and social networks proactively communicated scientific information about face masks and other health-promoting measures to their users. AI tools were also deployed in the search for COVID-19 therapeutics and vaccines.

As the pandemic recedes from view, it's time to chart a new path forward. Despite deep political polarization and legislative stalemates in many democracies, especially in the United States, our politics are open to a serious moment of reckoning with technology. The adoption of GDPR in Europe in 2018 presaged greater government involvement in regulating the tech sector. In Washington, DC, a bipartisan coalition opened hearings regarding antitrust action against a number of big tech companies just before the November 2020 election. And CCPA has stimulated potential action by Congress to enact federal privacy legislation across all fifty states. The AI arms race between China and the United States has produced commitments by many democracies to invest billions of dollars in AI research and education. Perhaps most tellingly, some tech CEOs are now calling openly for federal regulation on questions concerning data privacy, facial recognition, Section 230 of the Communications Decency Act, and the development of AI.

These are early indications that a new relationship between government and the tech sector is a real possibility. Grassroots efforts by researchers and tech workers are increasingly focused on achieving socially beneficial outcomes from technology. The AI researcher Joy Buolamwini founded the Algorithmic Justice League to call attention to the often-malign effects of algorithms on marginalized populations. Tech workers are organizing within their own companies to bring social protections to gig workers. In the wake of the livestreamed Christchurch massacre, nongovernmental organizations around the world banded together with governments to create the Christchurch Principles, an effort to promote social media governance that better supports democracy and human rights. Universities in the United States have launched a public interest technology consortium, creating new pathways for young people to work on addressing social problems with technology. It's the kind of development that kindred spirits of Aaron Swartz should be thrilled to see.

If we look beyond the United States and Europe to other industrialized democracies, we can find powerful sources of inspiration. In Taiwan, Audrey Tang, who serves as digital minister, is one example of how government can forge a better relationship with technology for its citizens. Tang, a child prodigy who learned to code at an early age, became a prominent open source software developer in Silicon Valley before she was twenty and spent six years working for Apple from Taiwan. In the aftermath of the Sunflower Movement, a student-led effort to protest trade agreements between Taiwan and China, a new government was swept into office. Tang, who had broadcast video to the internet of the student occupation of the Parliament Building, was invited to join the new government to bring about wide-ranging changes in digital policy. Tang sees democracy as a mechanism for peaceful resolution of competing interests and thinks that democratic institutions can be enhanced by digital technologies. Instead of seeing regulation as an intrusion into the free operation of the marketplace or an inherent constraint on technological innovation, she views government policy as an important partner for sustainable development and civic empowerment.

With this approach, she built a new infrastructure in Taiwan to support tech start-ups. She helped set up one of the most extensive, reliable, and high-speed internet systems in the world, available even to those in rural areas. She pioneered new ways of deploying digital tools for civic participation and economic development, such as the vTaiwan platform, which enables offline and online consultation and runs hackathons that coordinate feedback from citizens on budgets, policies, and other social issues. The platform brings together government ministries, elected representatives, scholars, experts, business leaders, civil society organizations, and citizens with the aim of increasing legitimacy in the results of government action. According to Tang, there are more than 5 million active members (of the country's 23 million inhabitants) on the vTaiwan crowdsourcing open platform.

More recently, Tang helped lead Taiwan's extremely successful COVID-19 strategy. By the end of 2020, Taiwan had recorded only 9 deaths and fewer than 1,000 total cases. Of course, Taiwan is a small country, so a better measure of Taiwan's success is deaths per 100,000 people. In the United States, more than 160 people per 100,000 have died, whereas in Taiwan the number is .04 per 100,000, this despite the fact that Taiwan is less than a hundred miles from mainland China, the epicenter of the pandemic, and more than 1 million Taiwanese citizens work in China. The most important element in Taiwan's success, according to many health experts, has been its systematic use of a digital health infrastructure that enables contact tracing and immediate data on patient visits to hospitals across the country. The irony was not lost on prominent US health experts. As Ezekiel J. Emanuel, Cathy Zhang, and Aaron Glickman from the Department of Medical Ethics and Health Policy at the University of Pennsylvania wrote, "Americans share every movement and sentiment with Facebook and Google, yet we seem reluctant to allow the Department of Health and Human Services to monitor patient encounters, as Taiwan does, to track disease and determine what medical tests and treatments to order."

Such trust in public health will not emerge overnight in the United States, even or perhaps especially in the face of the pandemic. But building government expertise in technology, especially in light of the growing calls by politicians and citizens alike to rein in the power of big tech, would be a step in the right direction.

REBOOTING THE SYSTEM

Creating an alternative future that can engage all of us in the tensions and trade-offs surfaced by new technologies requires progress on three fronts: cultivating a greater appreciation and understanding of ethical issues among technologists, checking corporate power, and empowering citizens and democratic institutions to govern technology instead of passively allowing technology and technologists to govern us.

TECHNOLOGISTS, DO NO HARM

This isn't the first time that citizens in a democracy have had to face up to the malicious deployment, misuse, or unintended harms of technological advances. It helps to remember that democratic societies have confronted similar challenges in the past and emerged with frameworks that have helped preserve the benefits of technology while diminishing the harms.

In medical research and clinical care, for example, regulation helped move the field from unregulated quackery and unprincipled experimentation on human subjects to institutionalized norms that protect individual rights and public safety, with government bodies providing oversight. The fields of biomedical research and health care provide some important lessons for a professional evolution that needs to happen among technologists.

Modern medical practice is typically considered to begin with taking the Hippocratic Oath, the world's oldest ethical code, dating back to Hippocrates, a physician in Greece in the fourth century BC. The oath goes far beyond the idea of "First, do no harm" and involves a promise to promote the interests of patients and honor the ideals of the medical profession. A recitation of the pledge is a widely practiced ritual at medical school commencement ceremonies. Though it is not a legal obligation and there are no enforcement mechanisms, it holds symbolic power as a time-honored ritual. It's an induction, as described by former surgeon general C. Everett Koop, into "an ethical tradition that transcends the legal vicissitudes of time and the fickleness of the law." Inspired by the oath, there have been contemporary calls for finance professionals and engineers to adopt their own versions of it.

It was a pair of twentieth-century incidents, however, that served as a catalyst for the institutional development of professional ethics in medicine. Both were triggered by public acknowledgement of enormous harms. The first was the 1910 publication of a report commissioned by the newly formed Carnegie Foundation for the Advancement of Teaching that documented in painstaking detail the enormous variation in training and medical education in North America. Its author, Abraham Flexner, visited hundreds of medical schools, and his report provided choice commentary about the "indescribably foul" and "wretched rooms" of medical schools that constitute "a plague on the nation" alongside the lax standards for practicing doctors. Within a decade of its publication, the report led to the establishment of minimum standards for medical education as well as national licensing exams and continuing professional education requirements, all to be authorized by state medical boards. Those medical boards remain essential components of the practice of medicine today, with power to accredit medical schools and sanction, or even withdraw, medical licenses for health care providers who violate state laws or professional codes of conduct.

The second was the global reckoning in the aftermath of World War II when twenty-three doctors stood trial at Nuremberg, Germany, for alleged crimes they had committed by conducting torturous and murderous experiments on Jewish prisoners. From the trial emerged the 1947 Nuremberg Code, which contains ten principles intended to govern all medical research and experimentation that involves humans. Foremost among its principles is the adoption of informed and voluntary consent of every subject. This represented a tectonic shift in medical research, which had been guided for decades by a principle of medical utilitarianism, in which the prospect of substantial social benefits outweighed the potential harm or suffering of individuals. The Nuremberg Code protects the interests of patients and subjects of experiments and drug trials, and it has served as the basis of numerous human rights laws and medical ethics codes around the world. It led to the creation of a new field of scholarly inquiry, bioethics, that is now ubiquitous in medical schools and gave rise to ethics committees attached to hospitals that provide guidance on difficult cases. A few decades later, the code guided the response in the United States to the shocking revelation of the decade-long Tuskegee experiment, in which doctors withheld lifesaving treatment for syphilis from a subset of six hundred African American men in order to study the natural evolution of the disease. Congressional hearings on the experiment led to a National Commission for the Protection of Human Subjects of Biomedical and Behavioral Research, whose 1979 Belmont Report led in time to the adoption of the "Common Rule" in 1991. That rule established institutional review boards for all research involving human subjects, a rigorous evaluation that requires an ethical assessment of the benefits and risks of the proposed research and, except in very limited circumstances, enforces informed consent.

Today, medical practice and research are structured by a thick institutional web of professional norms, legal codes, state licensure bodies, federal agencies, and human rights doctrines. The Food and Drug Administration must authorize drug trials, conducted according to rigorous and uniform standards, before any pharmaceutical product can be released. The World Health Organization and the European Medicines Agency play analogous roles outside the United States. It's a well-established practice, enforced by the 1996 Health Insurance Portability and Accountability Act (HIPAA), that data about your individual health are held to strict privacy standards. The government frequently establishes national commissions to make recommendations about controversial issues regarding emerging technologies, such as the Presidential Commission for the Study of Bioethical Issues during the Obama administration, which studied cutting-edge topics of bioengineering and the use of fetal tissue and stem cells in research. The result is that individuals who interact with the medical establishment, whether through clinical care or research, can trust that doctors are held to common standards of training and the interests of the patient are foremost. They can trust that there are serious efforts to test and regulate the availability of pharmaceuticals, so that drugs available over the counter or via prescription have been vetted by professionals. In the span of a century, the experience of being a doctor or a patient was transformed through the advancement and institutionalization of medical ethics.

The experience in medicine does not provide a complete playbook for what's needed for technologists. Yet it offers a lesson about what's possible with greater public awareness and informed public policy. We should not be especially surprised that analogous efforts are yet to be seen in the technology sector. The reforms in medicine took decades, often in response to scandals that aroused public outrage. Computer science is a much younger field, and the rise of digital technology and Silicon Valley is more recent still. As public awareness has grown about the dangers of technology, the time is ripe for organized efforts to strengthen and institutionalize the professional ethics of software engineers and computer programmers.

The main professional society for computer scientists, the Association for Computing Machinery (ACM), has already been exploring these ideas. In the late 1990s, the ACM established a task force to examine whether software engineers should be licensed. In some engineering fields, the National Society of Professional Engineers accredits university degree programs and creates licensure exams for those who work on safety-critical systems, such as bridges and building construction. If a bridge or building collapses because of mistakes made by professional engineers, companies are legally liable for damages and individuals can have their licenses to practice revoked. Software engineers bear almost no such responsibility. A clause, for example, buried in the license terms for Microsoft Excel provides a broad shield to Microsoft and its engineers from legal liability if it should turn out that the spreadsheet software contains coding errors that lead to miscalculations. Similarly, purchasers of buggy software or apps typically have little legal recourse. Though it's simply not realistic to expect all software to be error free before it is released, it is reasonable to expect that certain processes and best practices be followed to try to determine and mitigate the negative consequences of software as part of the development process. Such processes should be applied on a regular basis—for example, in the development of every new version or update of the software—to allow for continual reevaluation.

The ACM task force ultimately rejected the idea of licensing requirements. "The licensing of software engineers as PEs [professional engineers]," the committee concluded, "at best would be ignored and at worst would be damaging to our field. It would have no or negligible effect on safety." A sister organization, the Institute of Electrical and Electronics Engineers (IEEE), had a different view and sought to create a professional licensing exam. After several years of development, the exam was launched in a voluntary phase in 2013. But that effort also foundered, due mainly, it seems, to lack of uptake from graduating students, as the licensing is not required for industrial practice. By 2018, the exam had been administered five times to a total of eighty-one candidates. The IEEE shuttered the project. Many in the computer science community continue to resist professional licensure. Requiring a software engineer, for example, to receive a degree from an accredited university program as a prerequisite to licensing would eliminate the possibility of a future Bill Gates or Mark Zuckerberg, two university dropouts whose coding skills were the basis of the companies they started.

Despite the difficulties with licensure, the ACM and IEEE Computer Society collaborated on an effort to produce a Software Engineering Code of Ethics, which was first released in 1997. The code, which has been updated a number of times since, includes a number of laudable, if airy, principles. Most notably, there is little in the way of significant consequences for violating the code. A verifiable violation would result in expulsion from ACM for sure, but membership in the organization is entirely voluntary to begin with and is not a requirement for being a software engineer.

A robust infusion of ethics into the culture of software engineering requires efforts on three major fronts. First, we need to broaden the practice of what is called "value-based design." The idea is to push a discussion about values, and especially about value trade-offs, into the early design phase of any technology. Raising questions about ethics should not be viewed simply as a matter of legal compliance. Value-based design reflects the awareness that technologies are not value neutral. As we have seen, technologies embed within them certain choices about values such as privacy and safety. Creating technology teams inside companies that bring different skill sets from engineering, social science, and ethics will help frame and pose design choices in the development of any technology. As Jack Dorsey lamented about his early hiring decisions at Twitter, he wished he had hired social scientists who could understand and model the effect on human behavior of creating a "like" button and counter for how many likes any post has received.

Second, professional bodies such as the ACM and IEEE should inject a dose of steroids into a renewed conversation about professional norms, an enhanced code of ethics, and potential licensure. The goal is to bring about a greater sense of professional identity so that norms that guide the work of technologists will serve as a mechanism for policing bad behavior that sits outside the official sanction of the law.

Take the example of the Chinese biologist He Jiankui. In 2018, He used a powerful new gene-editing procedure called CRISPR to edit the genome of twin girls while they were still just embryos. CRISPR's chief discoverers, Jennifer Doudna and Emmanuelle Charpentier, who won the Nobel Prize in Chemistry in 2020 for their breakthrough, recognized its potential for misuse. Doudna was motivated by a nightmare that haunted her: "Suppose somebody like Hitler had access to this—we can only imagine the kind of horrible uses he could put it to." To prevent such outcomes and to preserve public trust in the powerful new discovery, she led an effort to create a moratorium on its clinical use on humans. Professional scientific societies endorsed the temporary ban as a norm for practicing respectable science. But He disregarded the ban. When he announced his action to the world at a scientific conference, the reaction was severe. He was fired from the university where he worked and disinvited from scientific meetings, and his research papers were not accepted by any journal. He became a pariah in the scientific community. In late 2019, Chinese authorities sentenced him to three years in prison for his scientific malfeasance.

He's case shows the importance of a strong sense of ethical norms in a professional community. Such norms, coupled with consequences for their violation, create a greater sense of responsibility among practitioners not just to do what is legal but to do what the scientific community and broader public consider ethical. This expectation can serve as a powerful buffer for working with new technologies for which there has not yet been time for the thoughtful deliberation needed for regulation.

Technologists are only just beginning to take seriously the project of developing professional norms as strong as those in biomedical research. A number of groups, such as the European Union's High-Level Expert Group on Artificial Intelligence, have called for responsible publication guidelines that would identify when researchers should limit the release of new AI models. In the absence of such guidelines, we'll continue to see controversy. As in the case of CRISPR, progress will require that the field's most prominent scientists take a leading role. Yet Google's firing in 2020 of Timnit Gebru, a leading AI ethics researcher, raises questions about the willingness of tech companies to accept the ethical critiques of those within their own ranks.

Equally important is the development of norms that sanction rule breakers. One ethically dubious strain of AI research, for example, involves the deployment of facial recognition tools to make predictions about various forms of human identity or behavior such as homosexuality or criminal tendencies. Such efforts appear to be a modern revival of physiognomy, the discredited scientific practice of inferring inner traits from outside appearances. An Israeli company founded in 2014, Faception, claims to reveal personality traits based on facial images, including the use of "proprietary classifiers" that identify extroverts, high-IQ people, professional poker players, and threats. And in early 2020, several academic researchers announced that they would publish a paper titled "A Deep Neural Network Model to Predict Criminality Using Image Processing" in a forthcoming book. There was also a paper published in *Journal of Big Data* by Mahdi Hashemi and Margeret Hall called "Criminal Tendency Detection from Facial Images and the Gender Bias Effect" that claimed to distinguish likely from unlikely criminals on the basis of "the shape of the face, eyebrows, top of the eye, pupils, nostrils, and lips."

In a development that shows the beginnings of stronger professional norms among technologists, more than two thousand academics and industry researchers signed an open letter that called upon the articles to be retracted and for journals to refrain from publishing any similar research because, they wrote, such work is infected by racial bias and contributes to racialized practices of incarceration. The effort succeeded: Hashemi and Hall retracted their paper, and Hall publicly disavowed the work, saying she supported the norms laid out in the open letter.

Finally, we need an overhaul of the way we teach young software engineers and aspiring tech entrepreneurs. Just as young biology and medical school students take courses in bioethics and draw upon a deep well of research in this area, so, too, should computer science departments develop ambitious and interdisciplinary new courses that draw upon a young but growing base of scholarship on the ethical and social dimensions of technology.

The work we are doing at Stanford, teaching a course that integrates computer science, social science, and ethics, is just one example of a budding revolution that's under way at institutions across the United States and around the world. Teaching computer scientists is no longer the domain only of computer science professors, as voices from across the disciplines offer unique perspectives that technologists will benefit from professionally. The idea is simple: to create a new generation of civic-minded technologists and policy makers. Modeled on the emergence of public interest law—an innovation that transformed legal education with the goal of preparing young lawyers to work in nonprofit organizations and the public sector—the ambition is big. It's a world in which the civic-mindedness of technologists such as Aaron Swartz is no longer considered exceptional.

In transforming how we teach technologists, we must also pay attention to whom we are teaching. Since new technologies encode the needs, perspectives, and values of those who create (and fund) them, it's no surprise that the lack of diversity in tech companies is part of the problem. It helps explain algorithmic bias, an inattention to the ways that surveillance has been misused, a lack of concern for the distributional harms caused by automation, and the proliferation of hate speech online. The movement afoot among companies and funders to recruit, support, and retain a diverse field in technology is long overdue. But we also need to work in education, beginning in elementary and secondary school through university, to build a more diverse pipeline into tech. There's no substitute for the inclusion of diverse perspectives if we want to grapple seriously with the competing values at stake in the design of new technologies.

NEW FORMS OF RESISTANCE
TO CORPORATE POWER

Although she lacks the corporate power and personal wealth of the major tech company CEOs, Margrethe Vestager, the public face of Europe's effort to check the power of the major US technology companies, has been their most serious challenger to date. Under her leadership as the European Union's Commissioner for Competition, the EU levied fines against Google for abusing its dominance in search, Apple and Amazon for unpaid taxes, and Facebook for misleading regulators about its purchase of WhatsApp. Her regulatory actions spurred investigations and fines around the globe, as governments in Canada, Taiwan, Brazil, and India (among others) have also taken aim at the anti-competitive behavior of major American firms. She continues to press forward with even more aggressive action, including proposed new rules that would prevent the platforms from giving their own products better treatment over those of rivals. This could have a direct impact on the search results Google displays and the products Amazon promotes.

Vestager's approach is an attempt to curtail the power of the small number of companies that exercise disproportionate and unchecked influence on the way that technologies affect our society. Unsurprisingly, it was not popular in Silicon Valley and until recently hadn't found many fans in the United States. Many US politicians and regulators have accepted the argument that the big tech companies have earned their dominance—that it is a reflection of the high-quality services they provide and nothing more. Even President Obama maligned Vestager's investigations into big tech as sour grapes. In an interview in 2015, he said, "their service providers who, you know, can't compete with ours—are essentially trying to set up some roadblocks for our companies to operate effectively there."

But in recent years, the terrain has been changing fast. Both citizens and politicians are concerned about the untrammeled power and market domination of big tech. And there is increasing recognition that their power isn't only a result of their high-quality products. Instead, their dominance reflects unique features of information technology—"network effects" in which goods or services become more valuable as more people use them—and the decidedly hostile view toward regulation that enabled the growth of the information economy in the 1990s but failed to put any meaningful constraints into place.

US regulators are finally getting into the game and playing catch-up to Vestager and her EU colleagues. In late 2020, a flood of major lawsuits was filed in US courts against the big tech companies. The Federal Trade Commission and forty-eight US state attorneys general took aim at Facebook, arguing that the company had achieved its dominance by buying or burying its rivals, thus limiting consumers' choices and access to privacy protections. The FTC complaint cites internal company emails, many from Mark Zuckerberg himself, that strongly hint at anti-competitive strategies through deploying its network power, using its service for early detection of significant competitors and then acquiring any real threats. "It is better to buy than compete," wrote Zuckerberg in a June 2008 email.

Governments have more than Facebook in their sights. Attorneys general from thirty-five states filed suit against Google one week after the FTC complaint against Facebook, accusing the company of using anti-competitive practices to maintain its monopoly on search and search advertising. The lawsuit was explosive, revealing previously hidden deals with companies such as Apple to reinforce Google's dominant position in the marketplace. A related lawsuit alleges that Google and Facebook agreed to coordinate with each other to manipulate the market for online advertising. Although the cases won't be resolved anytime soon, they are an opening salvo in the battle to rein in the dominant market players.

The major US tech companies can no longer ignore this shifting landscape. Publicly, many of them are embracing the role of government as regulator, welcoming public deliberation about new laws and policies—all the while knowing that partisan polarization and gridlock will make change extremely unlikely. But just in case, behind the scenes, they are fighting these lawsuits and lobbying furiously to ensure that any new laws and regulations do not harm their bottom lines or dominant market positions.

One also hears corporate leaders delivering a stream of mea culpas, a recognition that perhaps they have not always operated with an eye to the public interest. In one particularly revealing version of this apology tour, Jack Dorsey, the CEO of Twitter, shared his view that Twitter's leaders had been unprepared for the ways in which the platform might be harnessed to do harm: "If we were to redo anything, it would just be to really look at some of the problems that we were assuming that we would face and make sure that we have the right skills and not assume that product managers and designers and engineers have those skill sets." Along with such self-reflections, we often hear a commitment to turn over a new leaf. In the wake of the Cambridge Analytica scandal, Facebook took out a full-page ad in several major newspapers in the United States and United Kingdom with a statement from Mark Zuckerberg: "We have a responsibility to protect your information. If we can't, we don't deserve it."

The commitment of leading companies to do better is a welcome development. It is taking concrete form in emerging perspectives on responsible innovation and new forms of corporate organization. Nicole Wong, a former vice president and deputy general counsel of Google—she was called "the Decider" by her colleagues—argues for taking a slower and more reflective stance in technology innovation. She is challenging her colleagues in tech to think about the "world that we actually ought to be *trying* to build" rather than designing for the world we are in now.

But all of this is far from enough. Because at its heart, the claim that companies will pursue a new "north star"—one that balances profit seeking with social concerns—is one that demands we put our faith in corporate CEOs to look after our social welfare. And there simply is too little evidence that we can trust them to do so.

The Nobel Prize–winning economist Joseph Stiglitz compared the business elite to a "dieter who would rather do anything to lose weight than actually eat less." From Stiglitz's perspective, business leaders would prefer to do anything other than "fundamentally question the rules of the game—or even alter their own behavior to reduce the harm of the existing distorted, inefficient, and unfair rules."

The evidence of the failures of self-regulation is all around us. The enthusiasm for deregulation in the financial sector, which began in the

1970s, brought us the Great Recession of 2009. Even one of the world's greatest enthusiasts for deregulation, former Federal Reserve chairman Alan Greenspan, has admitted the error of his ways. A naive belief in the self-healing power of markets stopped regulators from intervening even as complex financial innovations and greater market concentration created ever-increasing systemic risk—until it was too late.

If we want to effectively govern new technologies, we must demand more than corporate CEOs seeing the light and pledging better behavior. Structural changes are necessary to ensure that existing incentives do not nudge CEOs toward profit seeking above all else. To achieve the outcomes we desire requires policy changes in how we approach markets so that there are checks on corporate power and monopolistic behavior. In this respect, Margrethe Vestager is heading in the right direction. And she has a growing chorus of voices behind her, including in the United States.

An agenda to limit the power of big tech companies has three key components. The first is addressing the huge power imbalance between companies and consumers when it comes to control over users' personal data. A far more aggressive commitment to a right to data protection, alongside government agencies capable of enforcing that right, should be the first critical check on corporate power.

Such data protection should not only include regulations for how users' data are used and require consent for their collection—principles already well outlined in the GDPR—but also provide ways for moving data across platforms, with privacy concerns in mind. If Facebook users have already invested time in connecting with hundreds of friends and uploading countless photos, it's unlikely that they will switch to a new social network, even if it has better features or adheres to policies they like more. It would simply be too much work to re-create their existing online social environment. Google's ill-fated Google+ social network is proof that even a well-funded competitor could not succeed in the current environment. Data portability would allow users to move their data, such as pictures and posts, to a new platform, and interoperability would guarantee that they could maintain their online experience, including their connections to friends who have moved to other networks. This would create a more competitive marketplace, where users aren't locked into one platform but can more easily move to another platform they feel does a better job of protecting their privacy or is more aligned with their values. Technically, it's a tall order but not impossible. In fact, something called the OpenSocial specification was developed in 2007 by a consortium of tech companies led by Google to try to create such interoperability for social networks. But the concept didn't get broad traction, as dominant players like Facebook saw no advantage to adopting it—again providing an example of why government regulations are more likely to bring about needed changes even when they are rejected by market forces.

The second component is giving greater voice in companies to those who are likely to be hurt by technological change. Many corporations continue to be governed by the idea of maximizing shareholder return, but there are alternatives to consider. A host of legislative proposals motivated by a vision of "stakeholder capitalism" (rather than shareholder capitalism) represents a meaningful next step to redefining corporate responsibility, giving workers greater power on corporate boards, and reducing the incentives for company directors to privilege short-term profits over long-term returns. These proposals go beyond simple expressions of CEOs' commitment to a broader group of stakeholders; they include concrete legislative mandates, for example, the creation of a new federal charter for companies that would require them to consider the interests of all stakeholders; or a requirement that employees directly elect 40 percent of the governing boards of large corporations; or the placing of significant restrictions on the ability of directors and officers to sell shares of a company they receive as equity, in an effort to diminish the unhealthy focus on short-term shareholder returns. Elizabeth Warren's Accountable Capitalism Act, introduced in 2017, provides just one example of the form this might take, and broader coalitions are forming around new proposals for corporate governance reform.

The third component is an assertive effort to constrain the market dominance of the major tech companies. This means cracking down on monopolistic behavior and restricting anti-competitive mergers and acquisitions. Most countries are already following the European Union's lead on this front, and US antitrust enforcement activities are finally under way. But the legal battles are likely to take years. And the history of antitrust regulation in high tech shows that it need not be necessary to break up the big players to achieve results. Rather, the simple threat of stronger antitrust action would help rein in some of their more extreme anti-competitive practices, allowing competitors to emerge.

Beginning in the late 1990s, the Department of Justice and numerous state attorneys general filed suit against Microsoft for anti-competitive behavior aimed at maintaining its dominance in the software industry. Though the suit ultimately did not result in Microsoft being split into two separate companies, as had originally been ordered by the judge presiding over the case, the specter of antitrust enforcement was sufficient to change some of the company's business practices. Surprisingly to many at the time, in 1997 Microsoft invested $150 million in its rival Apple, which would have otherwise run out of cash within months. Many observers have speculated that the real reason for that investment was to temper the claims of Microsoft's market monopoly. As history would prove, Apple would not only survive but in 2010 would become a more valuable company than Microsoft. Similarly, arguments have been made that Microsoft's worries about further antitrust actions softened its competitive stance in the marketplace more broadly, allowing new companies such as Google to emerge as formidable players.

GOVERNING TECHNOLOGY
BEFORE IT GOVERNS US

Despite our enthusiasm for the role of democracy in governing technology, our democratic institutions do not always inspire much hope. There have been far too many moments when politicians have demonstrated their ignorance of how new technologies work. We have watched parties on both the left and the right curry favor with leading tech companies, cognizant of their market power and political influence. And the polarization and legislative gridlock in so many democratic societies make it difficult—if not impossible—to have reasoned discussions about how best to balance competing values. But there was a time not long ago when the US government had a world-renowned scientific advisory body, which was copied by other countries.

Jack Gibbons, a folksy physicist from Tennessee, was the director for more than a decade of a little-known congressional agency, the Office of Technology Assessment. The OTA was born in 1972 at a time of growing public concern over pollution, nuclear energy, pesticides, and other hazards related to technological change. That was almost a decade after the publication of *Silent Spring*, the book that jump-started the environmental movement and focused public attention on the ubiquitous, fast-growing new technologies that promised great benefits and also posed great risks.

With the creation of the OTA, Congress recognized the urgency of effectively bridging the gap between technological expertise and political decision making. Members of Congress wanted the substantive background necessary to make difficult policy calls but didn't want to be dependent on information provided by lobbyists. Over nearly two decades, the OTA produced more than 750 reports on an extraordinary range of topics including the environment (acid rain, climate change), national security (technology transfer to China and bioterrorism), and social issues (workplace automation and how technology affects certain social groups). One distinctive feature of OTA reports, beyond their incisive technical analysis, was their commitment to providing a range of policy options without advocating for a specific one. That allowed policy makers to benefit from technical input and advice but make the difficult political choices themselves.

The agency was not shy about providing its technical perspective on highly controversial and politically consequential issues. In 1984, Ashton Carter, a young physicist who later went on to serve as secretary of defense between 2015 and 2017, authored a report on President Ronald Reagan's beloved space-based missile defense program (commonly known as "Star Wars"). He spoke plainly in concluding that "a perfect or near-perfect defense" against nuclear missiles represented an illusory goal that "should not serve as the basis of public expectations or national policy." The Pentagon was outraged and demanded retraction of the report. But an expert review of the report confirmed its conclusions, and two subsequent studies cast further doubt on the political wisdom and technical feasibility of Reagan's signature defense initiative.

Independent scientific judgment was on the chopping block when Newt Gingrich became the speaker of the House of Representatives in 1994. Congress's focus on budget cuts meant making some tough decisions to bring down its own spending, though Republican representative Amo Houghton desperately tried to save the OTA under the slogan "You don't cut the future." One observer called the dismantling of the agency a "stunning act of self-lobotomy." However, as the OTA's last director, Roger Herdman, indicated, the decision had been about far more than budget cuts: "There are those who said the Speaker didn't want an internal Congressional voice that had views on science and technology that might differ from his."

Gingrich's preferred model of the interaction of technical expertise and policy was what he called a "free-market" approach: members of Congress should take the initiative to engage individual scientists and inform themselves. That approach was, of course, unworkable and ineffective. And the free market for scientific expertise that Gingrich unleashed is at least part of what accounts for the intense politicization of science that we see today.

The OTA was never explicitly eliminated, it was just defunded. So it lives on today as a ghostly reminder of what is possible in the United States. Despite its zombie status, it spawned efforts in most European countries that are still in use. The Netherlands Office of Technology Assessment (NOTA) made a significant improvement on the US model by seeking to move beyond relying just on scientific experts. NOTA also incorporates citizen deliberation. Audrey Tang's efforts in vTaiwan aim for a similar form of citizenship empowerment.

The story of the OTA is a cautionary tale about the role of expertise:

though it may be possible to design institutions that provide authoritative information about highly technical issues to inform public debate, such bodies are politically vulnerable, especially when the facts they present prove inconvenient for powerful political actors. The very existence of the OTA and its demise underscore an even more important point: if we are unsatisfied with the capability of our democratic institutions to do the work of governing new technologies, it's because we let our institutions get this way.

It is up to us to ensure that democracy delivers regulatory frameworks and tech policies that promote socially beneficial uses of novel technologies while containing the harmful externalities that are hard to foresee and emerge only over time. Our task is not only about improving the policies on the issues we have explored in the book; it's also about transforming government so that it's better able to grapple with the issues that new technologies will present in the future. This requires a reboot of the policy-making process: a serious investment in bringing technologists in, educating policy makers as well as citizens about technology, and rethinking our approach to making smart choices about regulation.

We must bring technologists to the policy table so that our decisions reflect an understanding of the role of technology. As the Partnership for Public Service put it in a recent report, "Nearly every national priority depends on an accurate, thorough and contemporary understanding of how to use and leverage modern technology." Yet no one would argue that governments are good at attracting, deploying, and retaining tech talent. Democratic governments move slowly. They pay far too little to compete with major tech companies. And they are risk averse.

But the creation of "digital services" in both the United Kingdom and the United States that have been successful at recruiting strong technical talent from Silicon Valley and around the world proves that many technologists will enthusiastically serve in the public interest if given the opportunity to work on significant projects that can benefit millions of people. What this requires is a large-scale effort to recruit what former White House official Christopher Kirchhoff calls "tech teammates" through flexible mechanisms that can pull in top-flight talent with cutting-edge expertise and industry experience outside of the formal civil service channels. Just 6 percent of the federal workforce in the United States is under the age of thirty, and a wave of retirements looms on the horizon; this is a perfect moment to rethink how to recruit and retain new kinds of expertise to serve the public interest.

But the optimization mindset of technologists can also be a problem when they enter government. Their insights are critical to understanding what's at stake and what's possible, but their perspectives are only one input into a process in which policy makers must ultimately make trade-offs. So our second priority must be to ensure that our politicians are technologically literate and not just informed by the lobbyists who are paid to offer them a particular view. This will require the reinvention of the OTA in order to provide independent advice and a channel for citizen input on issues of science and technology policy. One priority for a re-vamped OTA would be to deliver its analyses in ways that are transparent and publicly accessible in order to facilitate a more open, inclusive, and deliberative process regarding tech policy issues. It also demands a more serious role for science and technology policy expertise in the Executive Branch. In 2021, President Biden elevated the national science advisor to the Cabinet for the first time. Biden also appointed Alondra Nelson, a social scientist with expertise at the intersection of technology, race, and inequality, as the deputy national science advisor. Her prominent role in the administration will help lay the groundwork for the high-quality democratic deliberation we urgently need.

Third, we need a fresh approach to regulation. Innovation in the tech sector is both disruptive and unpredictable; regulatory structures are rigid and reactive. Regulatory overhauls often lag technological change by years, playing catch-up as the accumulating evidence of harmful or unintended consequences becomes too difficult to ignore. The slow pace of regulation has been good for technology companies, giving them al-most free rein to experiment, test, and scale up without attention to the consequences. But what society needs is a more responsive approach to regulation that will enable us to try out new policy frameworks and learn about their effects before locking in a long-term strategy. Experts call this "adaptive regulation," and although it seems promising in principle, it can be very difficult to put into practice.

The United Kingdom and Taiwan have been leaders in experimenting with this new approach through "regulatory sandboxes." This is how they work: Government officials invite innovators to bring a proof of concept or proposal for the deployment of a technology in new and untested areas. If the officials approve, the existing body of law most relevant to the innovation at hand is "forked," a term taken from software development in which a new version of a program's code is created. In essence, the government provisionally grants the innovator permission to deploy the technology but also to devise regulations for the new system for a year. After a year, the innovators and officials reconvene to weigh the benefits and drawbacks of the new approach. The idea is to create living laboratories where regulators can observe the real-world effects of particular regulatory models. Fintech has been the major beneficiary of this approach, with firms able to test new offerings on consumers in real markets while regulators observe and evaluate their potential benefits and harms.

Even if we reboot the way the policy-making process works, our democratic institutions will remain an afterthought for those who are driving technological change unless citizens hold politicians accountable for the harmful effects of unregulated technologies. Though it is easy to point a finger at Mark Zuckerberg when Facebook mines your personal data, he is able to do so *legally* because politicians have set up the rules that way. The same goes for the proliferation of misinformation online. We can be as frustrated as we want at the content moderation policies of the internet platforms, but they are already doing far more than governments require them to do. And the effects of automation on jobs? Companies are just doing what's in their shareholders' interest. If people find themselves without transitional income support or new educational pathways to improve their skills, it's because politicians have not stepped up to the plate to deal with the consequences of automation.

The power of democracy is that it can help protect us from the outcomes we most want to avoid. You can thank government regulation for ensuring that the water coming out of your tap is drinkable, that your food won't make you sick, and that you can drive your car to work with a reasonable sense of safety on the road. But our politicians will only step up if their jobs depend on it. So although the marriage of hackers and capitalists is central to the story of our contemporary dilemmas, the failures of our democratic institutions to intervene on our behalf are also to blame. We all need to know where our elected representatives stand on the most important technological issues of the day, and we must be prepared to punish them at the ballot box when we don't like the results.

● ■ • ■ ●

Although we have focused squarely on the choices democracies need to make about how to govern their technological future, we cannot ignore the rise of China. China provides the world an alternative model of governance—authoritarian, ruthlessly efficient, and with a record of sustained economic growth. The country is also actively seeking digital supremacy by investing unmatched resources in AI, obtaining massive reams of personal data, stealing intellectual property from other countries, wielding its technological prowess to gain influence and access around the world, and seeking to imprint on the global rule-making bodies its own approach to digital authoritarianism.

As we balance the promise and peril of new technologies in our own societies, we will need to think about how our policy choices and regulatory approaches interact with—and in fact, can be actively coordinated with—other countries that share similar values. Though each country will make its own choices, we can't lose sight of the importance of forging common rules for the digital realm. Otherwise, we could end up in a situation in which China's preference for state control will displace our long-standing commitment to an open internet, robust competition, and set of meaningful protections for digital rights.

One common assumption about technology and geopolitics is that the world stage is a contest among China's digital authoritarianism, America's digital innovation, and Europe's focus on regulation. This dynamic is what drives many tech leaders to warn about the "Chinese alternative" if regulation were to stifle innovation. But the choice between innovation and regulation is a false dichotomy. Unless democracies assert greater collective voice over technology policy, the choice we face could be between global technology companies that do not place the interests of individuals or democracies first and the authoritarian model of technology governance offered by China.

We admit that it's a strange time to be mounting a defense of democracy and civic empowerment as the antidote to big tech's current predicaments. The public's faith in our governing institutions is at historic lows. Yet we must also remember that the distrust in democracy is partly a product of the rise of technologists. The recommendation systems and algorithmic curation of the private platforms that constitute the infrastructure of our digital public sphere have contributed to polarization and supercharged the spread of misinformation. And the tech industry has contributed to a winner-take-all economy, which has in turn widened wealth and income inequality, phenomena that social scientists have repeatedly demonstrated undermine confidence in democratic institutions.

We believe that democracy must be defended. Democracies are committed, at least in principle, to the noble and enduring values of individual freedom and equality. Democracy is itself a kind of technology, a design for social problem solving whose chief virtues are its defense of individual rights, empowerment of citizens' voices, and adaptability to ever-changing social conditions. Though it is fragile, with its current vulnerabilities on stark display, its roots go back centuries. It has proven resilient to any number of challenges in the past. The regulation of our technological future will be its next defining challenge.

ACKNOWLEDGMENTS

Writing a book is often a painful, arduous, and solitary undertaking. Writing this book has been the opposite: the result of one of the most rewarding and enjoyable professional collaborations of our lives. What began as a coffee outside the Stanford library to discuss the wild growth in the number of computer science majors became a yearlong effort to design a new class on ethics, policy, and technology. In the process, we learned as much from each other as did our students from us. After two years of the course, we asked ourselves whether we might write a book in addition to teaching together.

We owe thanks to our extraordinary agent, Elyse Cheney, who answered our cold email inquiry, took a chance on us, and guided us through the entire process of producing a book proposal. That led us to the exceptional Gail Winston, our editor at HarperCollins, whose many years of experience helped pare down our prose and instill a confidence that we could reach a broad audience. We also thank our copyeditor Lynn Anderson for saving three academics from many embarrassing mistakes.

We owe a special debt of gratitude to Hilary Cohen. Hilary has been an equal collaborator on the creation of the course since the very beginning. It is through our joint work in designing and teaching the course that we came to develop a common language and framework for tackling these issues on the technological frontier. Never have we worked with a recent college graduate with such poise, energy, vision, leadership, and brilliance. The course would not exist without her, and this book would not exist without the course.

Sam Nicholson came to our rescue at two crucial moments: first in the development of our book proposal and then at the final stages of writing the book itself. He helped us unlearn the common sins of scholarly writing. (Here's the problem with you academics: First you tell people you're going to do X, then you say, Look, I'm doing X, and finally you remind everyone, I just did X. Stop doing that!) And he taught us about the power of a quick anecdote or narrative flourish.

We had the pleasure of working with an amazing team of undergraduate and graduate student research assistants. Thank you to Adrian Liu, Janna Huang, Wren Elhai, Ben Esposito, Jessica Femenias, Isabella Garcia-Camargo, Gabriel Karger, Ananya Karthik, Anna-Sofia Lesiv, Jonathan Lipman, Mohit Mookim, Alessandra Maranca, Valeria Rincon, Rebecca Smalbach, Chase Small, Chloe Stowell, and Antigone Xenopoulous. Adrian and Janna, in particular, were with us from nearly start to finish, and neither the substance of the book nor the final product would be the same without them. Both are destined to be extraordinary scholars.

We thank the many colleagues and friends who read and offered comments on various portions of the book: Yuna Blajer de la Garza, Maria Clara Cobo, Joshua Cohen, Deep Ganguli, Sharad Goel, Julia Greenberg, Andrew Han, Daphne Keller, Jennifer King, Sam King, Karen Levy, Larissa MacFarquhar, Nate Persily, Sarah Richards, Marietje Schaake, Rebecca Smalbach, Henry Timms, Leif Wenar, and Erin Woo.

We thank Anna Wiener, the author of *Uncanny Valley*, for suggesting the title of our book.

Additionally, we thank the programs and people at Stanford University, as well as the supporters of our work, that helped to make our collaboration possible: the School of Humanities and Sciences, the School of Engineering, the Center on Philanthropy and Civil Society, the Institute for Human Centered Artificial Intelligence, the Center for Ethics in Society, the Continuing Studies Program, Nemil Dalal, David Siegel, Graham and Cristina Spencer, Roy Bahat, and Lisa Wehden. We owe a special debt of gratitude to the hundreds of students at Stanford who discussed these issues with us, to the dozens of teaching assistants who provided ongoing counsel, and to the numerous industry professionals who participated in our evening classes run in partnership with Bloomberg Beta. We learned a great deal and honed much of our own thinking about how to weigh these complex trade-offs in conversation with all of them.

Finally, we acknowledge our most important supporters.

Rob thanks Heather Kirkpatrick, who is optimal in every possible way.

Mehran thanks Heather Sahami, who always keeps focus on what truly matters.

Jeremy thanks Rachel Gibson, who demonstrates every day the love and compassion that humans offer but robots never will.

NOTES

PREFACE

XII   former president of the university: George Packer, "Change the World: Silicon Valley Transfers Its Slogans—and Its Money—to the Realm of Politics," *New Yorker*, May 27, 2013.

INTRODUCTION

XVII   "I got thirty": Joshua Browder, interview by Antigone Xenopoulos, 2018.

XVII   challenged more than 160,000: Aamna Mohdin and Ananya Bhattacharya, "An AI-Powered Chatbot Has Overturned 160,000 Parking Tickets in London and New York," Quartz, June 29, 2016, https://qz.com/719888/an-ai-powered -chatbot-has-overturned-160000-parking-tickets-in-london-and-new-york/.

XVIII   Browder may have been responding: Elisha Chauhan, "Councils to Rake in a Crazy £900m from Parking This Year," *Sun*, July 16, 2018, https://www .thesun.co.uk/motors/6790002/council-parking-fine-earnings-total-profits -westminster/.

XVIII   £9 billion backlog: Hayley Dixon, "Councils to Make Record £1 Billion from Parking Charges," *Telegraph*, June 29, 2019, https://www.telegraph.co.uk/news /2019/06/28/councils-make-record-1-billion-parking-charges/.

XIX   "I would like to": Joshua Browder, interview by Antigone Xenopoulos, 2018.

XX   "turns out to consist": Aaron Swartz, "Stanford: Day 11," *Raw Thought* (blog), October 2, 2004, http://www.aaronsw.com/weblog/001428.

XX   Swartz became a young millionaire: Kathleen Elkins, "The First Thing Alexis Ohanian Bought After He Sold Reddit for Millions at Age 23," CNBC, July 25, 2018, https://www.cnbc.com/2018/07/25/the-1st-thing-alexis-ohanian-bought -after-he-sold-reddit-for-millions.html.

XX   Reddit is today: Julia Boorstin, "Reddit Raised $300 Million at a $3 Billion Valuation—Now It's Ready to Take on Facebook and Google," CNBC, February 11, 2019, https://www.cnbc.com/2019/02/11/reddit-raises-300-million-at -3-billion-valuation.html.

XXI "Information is power": Aaron Swartz, "Guerilla Open Access Manifesto," Archive .org, July 2008, https://archive.org/details/GuerillaOpenAccessManifesto /mode/2up.

XXIII "At most 'technology' conferences": Aaron Swartz, "Wikimedia at the Crossroads," *Raw Thought* (blog), August 31, 2006, http://www.aaronsw.com/weblog/wikiroads.

XXIII "Aaron Swartz this is for you": Larissa MacFarquhar, "The Darker Side of Aaron Swartz," *New Yorker*, March 11, 2013, https://www.newyorker.com/magazine/2013 /03/11/requiem-for-a-dream.

XXX total liberty for the wolves: Isaiah Berlin, *The Crooked Timber of Humanity: Chapters in the History of Ideas*, edited by Henry Hardy, 2nd ed. (Princeton: Princeton University Press, 2013), 12–13.

CHAPTER 1: The Imperfections of the Optimization Mindset

4 a huge sum: Devin Leonard, *Neither Snow nor Rain: A History of the United States Postal Service* (New York: Grove Press, 2016), 85.

6 acquire Netflix for $50 million: Reed Hastings and Erin Meyer, *No Rules Rules: Netflix and the Culture of Reinvention* (New York: Penguin, 2020); GQ Staff, "The Tale of How Blockbuster Turned Down an Offer to Buy Netflix for Just $50M," *GQ*, September 19, 2019, https://www.gq.com.au/entertainment/film-tv/the -tale-of-how-blockbuster-turned-down-an-offer-to-buy-netflix-for-just-50m /news-story/72a55db245e4d7f70f099ef6a0ea2ad9.

7 "I started wondering": VICE Staff, "This Man Thinks He Never Has to Eat Again," *VICE*, March 13, 2013, https://www.vice.com/en/article/pgxn8z/this-man-thinks -he-never-has-to-eat-again.

7 "In my own life": Robert Rhinehart, "How I Stopped Eating Food," Mostly Harmless, February 13, 2013, https://web.archive.org/web/20200129143618 /https://www.robrhinehart.com/?p=298.

8 "It's really the labor": Lizzie Widdicombe, "The End of Food," *New Yorker*, May 5, 2014, https://www.newyorker.com/magazine/2014/05/12/the-end-of-food.

9 "stultifying utilitarianism": Farhad Manjoo, "The Soylent Revolution Will Not Be Pleasurable," *New York Times*, May 28, 2014, https://www.nytimes.com /2014/05/29/technology/personaltech/the-soylent-revolution-will-not-be -pleasurable.html.

9 "Imagine a meal": Sam Sifton, "The Taste That Doesn't Really Satisfy," *New York Times*, May 24, 2015, https://www.nytimes.com/2015/05/25/technology/the -taste-that-doesnt-really-satisfy.html.

10 "The ideas of economists": John Maynard Keynes, *The Collected Writings of John

Maynard Keynes, ed. Elizabeth Johnson and Donald Moggridge, vol. 7, *The General Theory* (London: Cambridge University Press, 1978), 383.

11   "tool for solving": Thomas H. Cormen et al., *Introduction to Algorithms*, 3rd ed. (Cambridge, MA: MIT Press, 2009), 5.

12   George Dantzig: George B. Dantzig, "Linear Programming," *Operations Research* 50, no. 1 (February 2002): 42–47, https://doi.org/10.1287/opre.50.1.42.17798.

14   algorithmic insights as a form of wisdom: Brian Christian and Tom Griffiths, *Algorithms to Live By: The Computer Science of Human Decisions* (New York: Henry Holt, 2016).

17   "Google should not be in the business of war": Scott Shane and Daisuke Wakabayashi, "'The Business of War': Google Employees Protest Work for the Pentagon," *New York Times*, April 4, 2018, https://www.nytimes.com/2018/04/04/technology/google-letter-ceo-pentagon-project.html.

18   Andrew Bosworth revealed in an internal memo: Ryan Mac, Charlie Warzel, and Alex Kantrowitz, "Growth at Any Cost: Top Facebook Executive Defended Data Collection in 2016 Memo—and Warned That Facebook Could Get People Killed," BuzzFeed News, March 29, 2018, https://www.buzzfeednews.com/article/ryanmac/growth-at-any-cost-top-facebook-executive-defended-data.

CHAPTER 2: The Problematic Marriage of Hackers and Venture Capitalists

25   "Governments of the Industrial World": John Perry Barlow, "A Declaration of the Independence of Cyberspace," Electric Frontier Foundation, February 6, 1996, https://www.eff.org/cyberspace-independence.

26   "we witnessed (and benefited from)": Udayan Gupta, "Done Deals: Venture Capitalists Tell Their Story: Featured HBS John Doerr," Working Knowledge, December 4, 2000, https://hbswk.hbs.edu/archive/done-deals-venture-capitalists-tell-their-story-featured-hbs-john-doerr.

27   he would later offer an apology: Will Sturgeon, "'It Was All My Fault': VC Says Sorry for Dot-Com Boom and Bust," ZDNet, July 16, 2001, https://www.zdnet.com/article/it-was-all-my-fault-vc-says-sorry-for-dot-com-boom-and-bust/.

28   "OKRs have helped": John Doerr, *Measure What Matters: How Google, Bono, and the Gates Foundation Rock the World with OKRs* (New York: Penguin, 2018), xii.

28   The countercultural notion: Reid Hoffman and Chris Yeh, *Blitzscaling: The Lightning-Fast Path to Building Massively Valuable Companies* (New York: HarperCollins, 2018).

28   "Competition," he says, "is for losers": Peter Thiel, "Competition Is for Losers

with Peter Thiel (How to Start a Startup 2014: 5)," Y Combinator, uploaded March 22, 2017, https://www.youtube.com/watch?v=3Fx5Q8xGU8k.

30   he quickly rose: David Shaw, not uncoincidentally, received a PhD in computer science from Stanford University and was a professor at Columbia before starting the company that bears his name.

31   "the key result": Doerr, *Measure What Matters*, 23.

32   "biggest bet in nineteen years": Ibid., 3.

32   "first PowerPoint slide": Ibid., 7.

32   "the marriage of Google": Ibid., 11.

33   "I think it's worked out": Ibid., xi.

33   "As Microsoft CEO Satya Nadella": Ibid., 161.

34   "Watch time, and only watch time": Ibid.

34   "For example, we made it": Ibid., 164.

34   "Like any management system": Ibid., 9.

34   "Goals Gone Wild": Lisa D. Ordóñez et al., "Goals Gone Wild: The Systematic Side Effects of Overprescribing Goal Setting," *Academy of Management Perspectives* 23, no. 1 (February 1, 2009): 6–16, https://doi.org/10.5465/amp.2009.3700 7999.

34   "Goals may cause": Doerr, *Measure What Matters*, 9.

36   "CEO Lee Iacocca": Ordóñez et al., "Goals Gone Wild," 4.

37   "The Social Responsibility of Business": Milton Friedman, "A Friedman Doctrine— The Social Responsibility of Business Is to Increase Its Profits," *New York Times*, September 13, 1970, https://www.nytimes.com/1970/09/13/archives/a -friedman-doctrine-the-social-responsibility-of-business-is-to.html.

37   He explicitly rejected: Ibid.

37   "There is one": Ibid.

38   "That goal is money": C. Wright Mills, *The Power Elite* (New York: Oxford University Press, 2000), 164.

38   "The biggest secret": Peter Thiel and Blake Masters, *Zero to One: Notes on Startups, or How to Build the Future* (New York: Crown Business, 2014), 86.

38   "it remains pretty rare": "Your Startup Has a 1.28% Chance of Becoming a Unicorn," CB Insights Research, May 25, 2015, https://www.cbinsights.com /research/unicorn-conversion-rate/.

39   Stanford itself got into: Ann Grimes, "Why Stanford Is Celebrating the Google IPO," *Wall Street Journal*, August 23, 2004, https://www.wsj.com/articles /SB109322052140798129.

39 venture fund investment exceeded $100 billion: Tom Nicholas, *VC: An American History* (Cambridge, MA: Harvard University Press, 2019), 268.

39 of all publicly traded companies founded after 1979: Will Gornall and Ilya A. Strebulaev, "The Economic Impact of Venture Capital: Evidence from Public Companies," Working Paper no. 3362 (Stanford: Stanford University Graduate School of Business, November 1, 2015), https://www.gsb.stanford.edu/faculty-research/working-papers/economic-impact-venture-capital-evidence-public-companies.

39 "To prioritize speed": Reid Hoffman, "7 Counterintuitive Rules for Growing Your Business Super-Fast," Medium, October 17, 2018, https://marker.medium.com/7-counterintuitive-rules-for-growing-your-business-super-fast-9dcdc2bfc649.

39 "We believe that the responsibilities": Hoffman and Yeh, *Blitzscaling*, 283.

39 "Recently we were asked": Jack Dorsey (@jack), "Recently we were asked a simple question: could we measure the 'health' of conversation on Twitter? This felt immediately tangible as it spoke to understanding a holistic system rather than just the problematic parts," Twitter, March 1, 2018, https://twitter.com/jack/status/969234282706169856.

40 "Investors are a simple-state machine": Elizabeth MacBride, "Why Venture Capital Doesn't Build the Things We Really Need," *MIT Technology Review*, June 17, 2020, https://www.technologyreview.com/2020/06/17/1003318/why-venture-capital-doesnt-build-the-things-we-really-need/.

41 "they all seem to be white, male, nerds": Sam Colt, "John Doerr: The Greatest Tech Entrepreneurs Are 'White, Male, Nerds,'" *Business Insider*, March 4, 2015, https://www.businessinsider.com/john-doerr-the-greatest-tech-entrepreneurs-are-white-male-nerds-2015-3.

41 only 2.3 percent of venture funding: Gené Teare, "Global VC Funding to Female Founders Dropped Dramatically This Year," Crunchbase News, December 21, 2020, https://news.crunchbase.com/news/global-vc-funding-to-female-founders/.

41 another CrunchBase report: Gené Teare, "EoY 2019 Diversity Report: 20 Percent of Newly Funded Startups in 2019 Have a Female Founder," Crunchbase News, January 21, 2020, https://news.crunchbase.com/news/eoy-2019-diversity-report-20-percent-of-newly-funded-startups-in-2019-have-a-female-founder/.

41 went to Black and Latinx founders: Crunchbase, "Crunchbase Diversity Spotlight 2020: Funding to Black & Latinx Founders," 2020, http://about.crunchbase.com/wp-content/uploads/2020/10/2020_crunchbase_diversity_report.pdf.

42 the tenth largest economy: Charles E. Eesley and William F. Miller, "Impact: Stanford University's Economic Impact via Innovation and Entrepreneurship," *Foundations and Trends in Entrepreneurship* 14, no. 2 (2018): 130–278, https://doi.org/10.1561/0300000074.

43   "On the back end": Marc Andreessen, "Why Software Is Eating the World," *Wall Street Journal*, August 20, 2011, https://online.wsj.com/article/SB10001424053 111903480904576512250915629460.html.

43   "most investments fail": Dave McClure, "99 VC Problems but a Batch Ain't 1: Why Portfolio Size Matters for Returns," Medium, August 31, 2015, https:// 500hats.com/99-vc-problems-but-a-batch-ain-t-one-why-portfolio-size -matters-for-returns-16cf556d4af0.

43   "If unicorns happen only": Ibid.

45   "Each batch of YC companies": "Investors," Y Combinator (website), June 2019, https://www.ycombinator.com/investors/.

45   "Since 2005, Y Combinator has funded": Y Combinator (website), https://www .ycombinator.com/.

45   Andreessen Horowitz created a separate fund: Meghan Kelly, "Andreessen-Horowitz to Give $50K to All Y Combinator Startups through Start Fund," VentureBeat, October 15, 2011, https://venturebeat.com/2011/10/14/andreessen -horowitz-to-give-50k-to-all-y-combinator-startups-through-start-fund/.

46   "Facebook and a variety": Megan Geuss, "Illinois Senator's Plan to Weaken Bio-metric Privacy Law Put on Hold," *Ars Technica*, May 27, 2016, https://arstechnica .com/tech-policy/2016/05/illinois-senators-plan-to-weaken-biometric-privacy -law-put-on-hold/.

46   Facebook denied the lobbying claim: Russell Brandom, "Facebook-Backed Law-makers Are Pushing to Gut Privacy Law," *Verge*, April 10, 2018, https://www .theverge.com/2018/4/10/17218756/facebook-biometric-privacy-lobbying-bipa -illinois.

46   the maximum $47 billion penalty: Bobby Allyn, "Judge: Facebook's $550 Million Settlement in Facial Recognition Case Is Not Enough," National Public Radio, July 17, 2020, https://www.npr.org/2020/07/17/892433132/judge-facebooks -550-million-settlement-in-facial-recognition-case-is-not-enough.

47   "It's $550 million": Ibid.

47   "If you sue us": Jared Bennett, "Saving Face: Facebook Wants Access Without Limits," Center for Public Integrity, July 31, 2017, https://publicintegrity.org /inequality-poverty-opportunity/saving-face-facebook-wants-access-without -limits/.

47   "Mark will be in Washington": Mike Allen, "Scoop: Mark Zuckerberg Re-turning to Capitol Hill," *Axios*, September 18, 2019, https://www.axios.com/mark -zuckerberg-capitol-hill-f75ba9fa-ca5d-4bab-9d58-40bcec96ff87.html.

47   Facebook and Amazon spent: Ryan Tracy, Chad Day, and Anthony DeBarros,

"Facebook and Amazon Boosted Lobbying Spending in 2020," *Wall Street Journal*, January 24, 2021, https://www.wsj.com/articles/facebook-and-amazon-boosted-lobbying-spending-in-2020-11611500400.

47 "These companies, because": Tony Romm, "Tech Giants Led by Amazon, Facebook and Google Spent Nearly Half a Billion on Lobbying over the Past Decade, New Data Shows," *Washington Post*, January 22, 2020, https://www.washingtonpost.com/technology/2020/01/22/amazon-facebook-google-lobbying-2019/.

48 "Washingtonization of Brussels": Adam Satariano and Matina Stevis-Gridneff, "Big Tech Turns Its Lobbyists Loose on Europe, Alarming Regulators," *New York Times*, December 14, 2020, https://www.nytimes.com/2020/12/14/technology/big-tech-lobbying-europe.html.

48 "As lawmakers, we will not": Matthew De Silva and Alison Griswold, "The California Senate Has Voted to End the Gig Economy as We Know It," Quartz, September 11, 2019, https://qz.com/1706754/california-senate-passes-ab5-to-turn-independent-contractors-into-employees/.

48 "app-based transportation (rideshare)": "California Proposition 22, App-Based Drivers as Contractors and Labor Policies Initiative (2020)," Ballotpedia, https://ballotpedia.org/California_Proposition_22,_App-Based_Drivers_as_Contractors_and_Labor_Policies_Initiative_(2020).

48 A consortium of tech companies: Kari Paul and Julia Carrie Wong, "California Passes Prop 22 in a Major Victory for Uber and Lyft," *Guardian*, November 4, 2020, https://www.theguardian.com/us-news/2020/nov/04/california-election-voters-prop-22-uber-lyft; Andrew J. Hawkins, "An Uber and Lyft Shutdown in California Looks Inevitable—Unless Voters Bail Them Out," *Verge*, August 16, 2020, https://www.theverge.com/2020/8/16/21370828/uber-lyft-california-shutdown-drivers-classify-ballot-prop-22.

49 "I doubt whether": Andrew J. Hawkins, "Uber and Lyft Had an Edge in the Prop 22 Fight: Their Apps," *Verge*, November 4, 2020, https://www.theverge.com/2020/11/4/21549760/uber-lyft-prop-22-win-vote-app-message-notifications.

49 "Up to 90%": Lyft, "What Is Prop 22 | California Drivers | Vote YES on Prop 22 | Rideshare | Benefits | Lyft," Lyft, October 8, 2020, https://www.youtube.com/watch?v=-7QJLgdQaf4.

49 "Unwarranted, concentrated economic power": Nancy Pelosi (@speakerpelosi), "Unwarranted, concentrated economic power in the hands of a few is dangerous to democracy—especially when digital platforms control content. The era of self-regulation is over," Twitter, June 3, 2019, https://twitter.com/speakerpelosi/status/1135698760397393921.

CHAPTER 3: The Winner-Take-All Race Between Disruption and Democracy

52 "One of the things": Maureen Dowd, "Peter Thiel, Trump's Tech Pal, Explains Himself," *New York Times*, January 11, 2017, https://www.nytimes.com/2017/01/11/fashion/peter-thiel-donald-trump-silicon-valley-technology-gawker.html.

52 systematically mapped the libertarian attitudes: David Broockman, Gregory Ferenstein, and Neil Malhotra, "Predispositions and the Political Behavior of American Economic Elites: Evidence from Technology Entrepreneurs," *American Journal of Political Science* 63, no. 1 (November 19, 2018): 212–33.

52 "Most members of Congress don't know": Kim Zetter, "Of Course Congress Is Clueless About Tech—It Killed Its Tutor," *Wired*, April 21, 2016, https://www.wired.com/2016/04/office-technology-assessment-congress-clueless-tech-killed-tutor/.

53 spent part of his honeymoon in Rome: Evan Osnos, "Can Mark Zuckerberg Fix Facebook Before It Breaks Democracy?," *New Yorker*, September 10, 2018, https://www.newyorker.com/magazine/2018/09/17/can-mark-zuckerberg-fix-facebook-before-it-breaks-democracy.

54 employment in sweatshops doubled: "History of Sweatshops: 1880–1940," National Museum of American History, https://americanhistory.si.edu/sweatshops/history-1880-1940.

55 "broad discretion in the investigation": Karen Bilodeau, "How the Triangle Shirtwaist Fire Changed Workers' Rights," *Maine Bar Journal* 26, no. 1 (Winter 2011): 43–44.

56 Though telegraphs were expensive: Richard Du Boff, "Business Demand and the Development of the Telegraph in the United States, 1844–1860," *Business History Review* 54, no. 4 (Winter 1980): 459–79.

57 "was able to charge monopoly prices": Tim Wu, "A Brief History of American Telecommunications Regulation," *Oxford International Encyclopedia of Legal History* 5 (2007): 95.

57 the government struggled: Ev Ehrlich, "A Brief History of Internet Regulation," Progressive Policy Institute, March 2014, https://www.progressivepolicy.org/wp-content/uploads/2014/03/2014.03-Ehrlich_A-Brief-History-of-Internet-Regulation1.pdf; Jonathan E. Nuechterlein and Philip J. Weiser, *Digital Crossroads: Telecommunications Law and Policy in the Internet Age*, 2nd ed. (Cambridge, MA: MIT Press, 2013).

58 "Finding good rules is not a one-time event": Paul M. Romer, *In the Wake of the Crisis: Leading Economists Reassess Economic Policy*, vol. 1 (Cambridge, MA: MIT Press, 2012), 96.

59 "Asymptotically, any finite tax code": Ibid.

59 In 1994, the Clinton administration: "Total Number of Websites," Internet Live Stats, January 4, 2021, https://www.internetlivestats.com/total -number-of-websites/; Elahe Izadi, "The White House's First Web Site Launched 20 Years Ago This Week. And It Was Amazing," *Washington Post*, October 21, 2014, https://www.washingtonpost.com/news/the-fix/wp/2014/10/21/the-white -houses-first-website-launched-20-years-ago-this-week-and-it-was-amazing/.

60 there were fewer than ten thousand websites: Wikipedia, s.v., "List of Websites Founded Before 1995," https://en.wikipedia.org/w/index.php?title=List_of _websites_founded_before_1995&oldid=997260381.

61 "an *oasis from regulation*": Ehrlich, "A Brief History of Internet Regulation."

62 Americans also pay much higher prices: Emily Stewart, "America's Monopoly Problem, Explained by Your Internet Bill," Vox, February 18, 2020, https:// www.vox.com/the-goods/2020/2/18/21126347/antitrust-monopolies-internet -telecommunications-cheerleading; Becky Chao and Claire Park, "The Cost of Connectivity 2020," New America Open Technology Institute, July 2020, https:// www.newamerica.org/oti/reports/cost-connectivity-2020/global-findings/.

63 lawsuit revealed the extent of Google's: United States District Court, "Case 4:20-Cv-00957," Court Listener, December 16, 2020, https://www.courtlistener .com/recap/gov.uscourts.txed.202878/gov.uscourts.txed.202878.1.0.pdf.

63 Tom Wheeler, FCC chairman: Tom Wheeler, "The Tragedy of Tech Compa- nies: Getting the Regulation They Want," The Brookings Institution, March 26, 2019, https://www.brookings.edu/blog/techtank/2019/03/26/the-tragedy-of -tech-companies-getting-the-regulation-they-want/.

64 "When [antitrust] laws": Bobby Allyn and Shannon Bond, "4 Key Takeaways from Washington's Big Tech Hearing on 'Monopoly Power,'" National Public Radio, July 30, 2020, https://www.npr.org/2020/07/30/896952403/4-key -takeaways-from-washingtons-big-tech-hearing-on-monopoly-power.

64 Amazon already controls: Katie Schoolov, "What It Would Take for Walmart to Catch Amazon in E-Commerce," CNBC, August 13, 2018, https://www.cnbc .com/2020/08/13/what-it-would-really-take-for-walmart-to-catch-amazon -in-e-commerce.html.

64 "dominant market share": Mark Gurman, "Apple's Cook Says App Store Opened 'Gate Wider' for Developers," Bloomberg, July 28, 2020, https:// www.bloomberg.com/news/articles/2020-07-29/apple-s-cook-says-app-store -opened-gate-wider-for-developers.

64 "Facebook is a successful": Mark Zuckerberg, "Testimony of Mark Zuckerberg, Facebook, Inc., Before the United States House of Representatives Committee on the Judiciary," July 9, 2020, https://docs.house.gov/meetings/JU/JU05 /20200729/110883/HHRG-116-JU05-Wstate-ZuckerbergM-20200729.pdf, 5.

65 "it is possible to love": Roger McNamee, "A Historic Antitrust Hearing in Congress Has Put Big Tech on Notice," *Guardian*, July 31, 2020, https://www.theguardian.com/commentisfree/2020/jul/31/big-tech-house-historic-antitrust-hearing-times-have-changed.

66 "most intelligence of the principles": Plato, *The Republic*, trans. Paul Shorey, vol. II (Books VI-X) (Cambridge, MA: Harvard University Press, 1942), 147.

66 "intoxicated by drinking": Ibid., 305–11.

66 economist Bryan Caplan: Bryan Caplan, *The Myth of the Rational Voter: Why Democracies Choose Bad Policies* (Princeton: Princeton University Press, 2008), 3.

66 the philosopher Jason Brennan tried: Jason Brennan, *Against Democracy* (Princeton: Princeton University Press, 2016).

68 "Experts can only offer alternatives": Thomas M. Nichols, *The Death of Expertise: The Campaign Against Established Knowledge and Why It Matters* (New York: Oxford University Press, 2017), 224.

69 would prefer to be governed by experts than by elected officials: Richard Wike et al., "Democracy Widely Supported, Little Backing for Rule by Strong Leader or Military," Global Attitudes & Trends, Pew Research Center, October 16, 2017, https://www.pewresearch.org/global/2017/10/16/democracy-widely-supported-little-backing-for-rule-by-strong-leader-or-military/.

69 "choosing leaders through free elections": Ian Bremmer, "Is Democracy Essential? Millennials Increasingly Aren't Sure—and That Should Concern Us All," NBC News, February 13, 2018, https://www.nbcnews.com/think/opinion/democracy-essential-millennials-increasingly-aren-t-sure-should-concern-us-ncna847476.

69 "There is no difficulty": John Stuart Mill, *Collected Works of John Stuart Mill*, ed. J. M. Robson, vol. 9, *Essays on Politics and Society* (Toronto: University of Toronto Press, 1977), 403.

70 Harvard professor Danielle Allen: Danielle S. Allen, *Our Declaration: A Reading of the Declaration of Independence in Defense of Equality* (New York: Liveright, 2014).

70 For Cohen, in debating questions of politics: Joshua Cohen, "Procedure and Substance in a Deliberative Democracy," in *Democracy and Difference: Contesting the Boundaries of the Political*, edited by Seyla Benhabib (Princeton: Princeton University Press, 1996), 95–119

71 "The general prosperity": Mill, *Essays on Politics and Society*, 404.

74 no democratic country has ever experienced a famine: Amartya Sen, *Poverty and Famines: An Essay on Entitlement and Deprivation* (Oxford: Oxford University Press, 1983).

75 "bad or incompetent rulers can be prevented": Karl R. Popper, *The Open Society and Its Enemies* (Princeton: Princeton University Press, 2013), 115.

75 "We should rather blame ourselves": Ibid., 120.

76 "a *summum bonum*": Judith Shklar, "The Liberalism of Fear," in *Liberalism and the Moral Life*, edited by Nancy L. Rosenblum (Cambridge, MA: Harvard University Press, 1989), 21–38.

76 "In a time of rapid technological change": Tom Wheeler, "Internet Capitalism Pits Fast Technology Against Slow Democracy," The Brookings Institution, May 6, 2019, https://www.brookings.edu/blog/techtank/2019/05/06/internet-capitalism-pits-fast-technology-against-slow-democracy/.

PART II: Disaggregating the Technologies

77 "the inventive genius of mankind": Albert Einstein, "The 1932 Disarmament Conference," *Nation*, September 4, 1931, repr. August 23, 2001, https://www.thenation.com/article/archive/1932-disarmament-conference-0/.

CHAPTER 4: Can Algorithmic Decision-Making Ever Be Fair?

79 "high bar for talent": Brad Stone, *The Everything Store: Jeff Bezos and the Age of Amazon* (New York: Little, Brown, 2013), 88.

80 "we can't afford to live by manual processes": Harry McCracken, "Meet the Woman Behind Amazon's Explosive Growth," *Fast Company*, April 11, 2019, https://www.fastcompany.com/90325624/yes-amazon-has-an-hr-chief-meet-beth-galetti.

80 "Everyone wanted this holy grail": Jeffrey Dastin, "Amazon Scraps Secret AI Recruiting Tool That Showed Bias Against Women," Reuters, October 10, 2018, https://www.reuters.com/article/us-amazon-com-jobs-automation-insight/amazon-scraps-secret-ai-recruiting-tool-that-showed-bias-against-women-idUSKCN1MK08G.

81 significant discrimination based on perceived race: Marianne Bertrand and Sendhil Mullainathan, "Are Emily and Greg More Employable than Lakisha and Jamal? A Field Experiment on Labor Market Discrimination," *American Economic Review* 94, no. 4 (2004): 991–1013.

81 downgraded applicants from women-only universities: Dastin, "Amazon Scraps Secret AI Recruiting Tool That Showed Bias Against Women."

86 labeled pictures of him and his girlfriend as "gorillas": Loren Grush, "Google Engineer Apologizes After Photos App Tags Two Black People as Gorillas," *Verge*, July 1, 2015, https://www.theverge.com/2015/7/1/8880363/google-apologizes-photos-app-tags-two-black-people-gorillas.

86 "This is 100% Not OK": Ibid.

86 eliminate all gorillas and chimpanzees from its image bank: Tom Simonite, "When It Comes to Gorillas, Google Photos Remains Blind," *Wired*, January 11, 2018, https://www.wired.com/story/when-it-comes-to-gorillas-google-photos-remains-blind/; James Vincent, "Google 'Fixed' Its Racist Algorithm by Removing Gorillas from Its Image-Labeling Tech," *Verge*, January 12, 2018, https://www.theverge.com/2018/1/12/16882408/google-racist-gorillas-photo-recognition-algorithm-ai.

88 highly likely to reoffend: Julia Angwin et al., "Machine Bias," ProPublica, May 23, 2016, https://www.propublica.org/article/machine-bias-risk-assessments-in-criminal-sentencing.

88 refused to divulge: Adam Liptak, "Sent to Prison by a Software Program's Secret Algorithms," *New York Times*, May 1, 2017, https://www.nytimes.com/2017/05/01/us/politics/sent-to-prison-by-a-software-programs-secret-algorithms.html.

88 they demanded an explanation: Angwin et al., "Machine Bias."

89 twenty-one distinct definitions of fairness: Arvind Narayanan, "Tutorial: 21 Fairness Definitions and Their Politics," uploaded to YouTube March 1, 2018, https://www.youtube.com/watch?vjIXIuYdnyyk.

89 several common specifications of fairness are incompatible: Alexandra Chouldechova, "Fair Prediction with Disparate Impact," *Big Data* 5, no. 2 (June 1, 2017): 153–63; Jon Kleinberg, Sendhil Mullainathan, and Manish Raghavan, "Inherent Trade-offs in the Fair Determination of Risk Scores," *Proceedings of Innovations in Theoretical Computer Science* 67, no. 43 (January 11, 2017): 1–23.

92 Two monkeys sat in adjacent cages: Sarah F. Brosnan and Frans B. M. de Waal, "Monkeys Reject Unequal Pay," *Nature* 425 (September 18, 2003): 297–99, https://doi.org/10.1038/nature01963.

94 The law eliminated cash bail: Vanessa Romo, "California Becomes First State to End Cash Bail After 40-Year Fight," National Public Radio, August 28, 2018, https://www.npr.org/2018/08/28/642795284/california-becomes-first-state-to-end-cash-bail.

95 "Thousands of sex offenders": Melody Gutierrez, "Bill to End Cash Bail Passes California Assembly amid Heavy Opposition," *San Francisco Chronicle*, August 20, 2018, https://www.sfchronicle.com/crime/article/California-legislation-to-end-cash-bail-loses-13169991.php.

95 "a transformational shift away": Romo, "California Becomes First State to End Cash Bail After 40-Year Fight."

95 42 percent fewer people could be jailed: Jon Kleinberg et al., "Human Decisions and Machine Predictions," *Quarterly Journal of Economics* 133, no. 1 (August 26, 2017): 237–93, https://doi.org/10.1093/qje/qjx032.

96 when it's hot outside: Anthony Heyes and Soodeh Saberian, "Temperature and Decisions: Evidence from 207,000 Court Cases," *American Economic Journal: Applied Economics* 11, no. 2 (April 19, 2017): 238–65, https://doi.org/10.1257 /app.20170223.

96 "not the model for pretrial justice": "ACLU of California Changes Position to Oppose Bail Reform Legislation," ACLU of Southern California, August 20, 2018, https://www.aclusocal.org/en/press-releases/aclu-california-changes-position -oppose-bail-reform-legislation.

96 "our rallying cry": Alexei Koseff, "Bill to Eliminate Bail Advanced Despite ACLU Defection," *Sacramento Bee*, August 20, 2018, https://www.sacbee.com /news/politics-government/capitol-alert/article217031860.html.

97 A highly publicized ProPublica investigation: Angwin et al., "Machine Bias."

97 were offered no-bail release: Tom Simonite, "Algorithms Should've Made Courts More Fair. What Went Wrong?," *Wired*, September 5, 2019, https://www.wired .com/story/algorithms-shouldve-made-courts-more-fair-what-went-wrong/.

98 "limitations and cautions": State v. Loomis, 881 N.W.2d 749 (Wisconsin 2016).

98 Cathy O'Neil: Cathy O'Neil, *Weapons of Math Destruction: How Big Data Increases Inequality and Threatens Democracy* (New York: Crown, 2016), 3.

98 "New Jim Code": Ruha Benjamin, *Race After Technology: Abolitionist Tools for the New Jim Code* (Medford, MA: Polity, 2019).

99 "We found zero relationship": Adam Bryant, "In Head-Hunting, Big Data May Not Be Such a Big Deal," *New York Times*, June 19, 2013, https://www.nytimes .com/2013/06/20/business/in-head-hunting-big-data-may-not-be-such-a-big -deal.html.

102 humans tend to unquestioningly accept the accuracy: Partnership on AI, "Report on Algorithmic Risk Assessment Tools in the U.S. Criminal Justice System," The Partnership on AI, 2019, https://www.partnershiponai .org/report-on-machine-learning-in-risk-assessment-tools-in-the-u-s-criminal -justice-system/.

102 judges are more likely to override: Alex Albright, "If You Give a Judge a Risk Score: Evidence from Kentucky Bail Decisions," The Little Data Set, September 3, 2019, https://thelittledataset.com/about_files/albright_judge_score.pdf, 1.

103 the US Air Force Academy experimented: Scott E. Carrell, Bruce I. Sacerdote, and James E. West, "From Natural Variation to Optimal Policy? The Importance of Endogenous Peer Group Formation," *Econometrica* 81, no. 3 (2013): 855–82, https://doi.org/10.3982/ECTA10168.

104 James Vacca, who was serving: Lauren Kirchner, "Algorithmic Decision Making and Accountability," Ethics, Technology & Public Policy, Stanford University,

August 16, 2017, https://ai.stanford.edu/users/sahami/ethicscasestudies/Algo
rithmicDecisionMaking.pdf.

105   "outsized influence": Rashida Richardson, "Confronting Black Boxes," AI Now Institute, December 4, 2019, https://ainowinstitute.org/ads-shadowreport-2019.pdf.

106   perhaps enabling greater progress: Jon Kleinberg et al., "Discrimination in the
Age of Algorithms," *Journal of Legal Analysis* 10 (2018): 113–74, https://academic
.oup.com/jla/article/doi/10.1093/jla/laz001/5476086.

107   automated tools for hiring decisions: Tom Simonite, "New York City Proposes
Regulating Algorithms Used in Hiring," *Wired*, January 8, 2021, https://www
.wired.com/story/new-york-city-proposes-regulating-algorithms-hiring/.

CHAPTER 5: What's Your Privacy Worth?

111   "showed up at my house": Sopan Deb and Natasha Singer, "Taylor Swift Said to
Use Facial Recognition to Identify Stalkers," *New York Times*, December 13, 2018,
https://www.nytimes.com/2018/12/13/arts/music/taylor-swift-facial-recognition
.html.

111   Enter ISM Connect: Gabrielle Canon, "How Taylor Swift Showed Us the Scary
Future of Facial Recognition," *Guardian*, February 15, 2019, http://www
.theguardian.com/technology/2019/feb/15/how-taylor-swift-showed-us-the
-scary-future-of-facial-recognition.

112   database of known Swift stalkers: Steve Knopper, "Why Taylor Swift Is Using
Facial Recognition at Concerts," *Rolling Stone*, December 13, 2018, https://www
.rollingstone.com/music/music-news/taylor-swift-facial-recognition-concerts
-768741/.

112   Baltimore is a case in point: Caroline Haskins, "Why Some Baltimore Residents
Are Lobbying to Bring Back Aerial Surveillance," The Outline, August 30, 2018,
https://theoutline.com/post/6070/why-some-baltimore-residents-are-lobbying
-to-bring-back-aerial-surveillance.

112   Joy Buolamwini: Gender Shades (website), http://gendershades.org/. See also Joy
Buolamwini and Timnit Gebru, "Gender Shades: Intersectional Accuracy Disparities in Commercial Gender Classification," *Proceedings of Machine Learning Research* 81 (2018): 1–15, http://proceedings.mlr.press/v81/buolamwini18a
/buolamwini18a.pdf.

115   billionaire John Catsimatidis: Kashmir Hill, "Before Clearview Became a Police
Tool, It Was a Secret Plaything of the Rich," *New York Times*, March 5, 2020,
https://www.nytimes.com/2020/03/05/technology/clearview-investors.html.

115   "surveillance capitalism": Shoshana Zuboff, *The Age of Surveillance Capitalism:*

The Fight for a Human Future at the New Frontier of Power (New York. PublicAffairs, 2019).

117 $130 billion in advertising revenue: J. Clement, "Google: Ad Revenue 2001–2018," Statista, 2020, https://www.statista.com/statistics/266249/advertising-revenue-of -google/.

118 "willingly accept legal terms": "2017 Global Mobile Consumer Survey: US Edition," Deloitte Touche Tohmatsu Limited, 2017, https://www2.deloitte.com /content/dam/Deloitte/us/Documents/technology-media-telecommunications /us-tmt-2017-global-mobile-consumer-survey-executive-summary.pdf.

118 "Specifically, when you share": "Facebook's Terms of Service," Facebook, https:// www.facebook.com/terms.php.

121 Bentham's own wishes: "Extract from Bentham's Will," Bentham Project, May 30, 1832, https://www.ucl.ac.uk/bentham-project/who-was-jeremy-bentham/auto -icon/extract-benthams-will.

121 his idea of a panopticon: Jeremy Bentham, *The Panopticon Writings*, ed. Miran Božovič (London: Verso, 1995).

121 a catalogue of the social benefits: Ibid., 31.

122 a "new mode of obtaining power": Ibid., 34.

122 F-House: Jonah Newman, "Stateville Prison Reopens Decrepit 'F-House' to Hold Inmates with COVID-19," Injustice Watch, May 12, 2020, https://www .injusticewatch.org/news/prisons-and-jails/2020/stateville-roundhouse-covid/.

122 The French philosopher Michel Foucault: Michel Foucault, *Discipline and Punish: The Birth of the Prison* (New York: Pantheon, 1977).

124 they installed two webcams: "Watching You Watching Bentham: The PanoptiCam," UCL News, March 17, 2015, https://www.ucl.ac.uk/news/2015 /mar/watching-you-watching-bentham-panopticam.

126 "It doesn't really bother me": Katherine Noyes, "Scott McNealy on Privacy: You Still Don't Have Any," *Computerworld*, June 25, 2015, https://www.computer world.com/article/2941055/scott-mcnealy-on-privacy-you-still-dont-have-any .html.

126 in 2017, Berners-Lee wrote: Tim Berners-Lee, "Three Challenges for the Web, According to Its Inventor," World Wide Web Foundation, March 12, 2017, https://webfoundation.org/2017/03/web-turns-28-letter/.

128 In 2014, Facebook bought WhatsApp: Henry Blodget, "Everyone Who Thinks Facebook Is Stupid to Buy WhatsApp for $19 Billion Should Think Again . . . ," *Business Insider*, February 20, 2014, https://www.businessinsider.com/why -facebook-buying-whatsapp-2014-2.

128   provide end-to-end encryption: Mark Zuckerberg, "A Privacy-Focused Vision for Social Networking," Facebook Newsroom, March 6, 2019, https://about.fb .com/news/2019/03/vision-for-social-networking/.

130   "GIC decided to release [135,000] records": Paul Ohm, "Broken Promises of Privacy: Responding to the Surprising Failure of Anonymization," *UCLA Law Review* 57 (2010): 1701–77, https://papers.ssrn.com/sol3/papers.cfm?abstract _id=1450006.

130   87 percent of all Americans: L. Sweeney, "Simple Demographics Often Identify People Uniquely" (Data Privacy Working Paper 3, Carnegie Mellon University, 2000).

130   In fact, the Data Privacy Lab: "How Unique Am I?," AboutMyInfo, https:// aboutmyinfo.org/identity.

131   Harvard professor Cynthia Dwork: Cynthia Dwork, "Differential Privacy," in *Automata, Languages and Programming*, edited by Michele Bugliesi et al. (Heidelberg, Germany: Springer, 2006), 1–12.

135   Cook refused to cooperate: Leander Kahney, "The FBI Wanted a Backdoor to the iPhone. Tim Cook Said No," *Wired*, April 16, 2019, https://www.wired.com /story/the-time-tim-cook-stood-his-ground-against-fbi/.

135   "We're not trying": Matt Burgess, "Google Got Rich from Your Data. DuckDuckGo Is Fighting Back," *Wired*, June 8, 2020, https://www.wired.co.uk/article /duckduckgo-android-choice-screen-search.

136   Google had acted illegally: Ibid.

137   Mounting evidence suggests that they don't: Alessandro Acquisti, Laura Brandimarte, and George Loewenstein, "Privacy and Human Behavior in the Age of Information," *Science* 347, no. 6221 (2015): 509–14, https://doi.org/10.1126 /science.aaa1465.

137   "36% of content": Yabing Liu et al., "Analyzing Facebook Privacy Settings: User Expectations vs. Reality," in *IMC '11: Proceedings of the 2011 ACM Internet Measurement Conference* (New York: Association for Computing Machinery, 2011), 61–70, https://dl.acm.org/doi/10.1145/2068816.2068823.

137   A colleague at Stanford, Susan Athey: Susan Athey, Christian Catalini, and Catherine Tucker, "The Digital Privacy Paradox: Small Money, Small Costs, Small Talk," Stanford Institute for Economic Policy Research, September 2017, https://siepr.stanford.edu/research/publications/digital-privacy-paradox-small -money-small-costs-small-talk.

139   broadly (more than 70 percent) concerned: Brooke Auxier, "How Americans See Digital Privacy Issues amid the COVID-19 Outbreak," Pew Research Center, May 4, 2020, https://www.pewresearch.org/fact-tank/2020/05/04/how -americans-see-digital-privacy-issues-amid-the-covid-19-outbreak/.

140 "cumulatively and holistically": Daniel J. Solove, "Introduction: Privacy Self-Management and the Consent Dilemma," *Harvard Law Review* 126 (2013): 1880–903, https://harvardlawreview.org/wp-content/uploads/pdfs/vol126_solove.pdf.

140 professor of law Daniel Solove: Solove, "Privacy Self-Management."

141 Apple and Google have built: Patrick Howell O'Neill, "How Apple and Google Are Tackling Their Covid Privacy Problem," *MIT Technology Review*, April 14, 2020, https://www.technologyreview.com/2020/04/14/999472/how-apple-and-google-are-tackling-their-covid-privacy-problem/.

142 Europe has always been more interested: Olivia B. Waxman, "The GDPR Is Just the Latest Example of Europe's Caution on Privacy Rights. That Outlook Has a Disturbing History," *Time*, May 24, 2018, https://time.com/5290043/nazi-history-eu-data-privacy-gdpr/.

143 Snowden's "comrade-in-arms": Simon Shuster, "E.U. Pushes for Stricter Data Protection After Snowden's NSA Revelations," *Time*, October 21, 2013, https://world.time.com/2013/10/21/e-u-pushes-for-stricter-data-protection-after-snowden-nsa-revelations/.

144 "protect an interest which is essential": General Data Protection Regulation, Regulation (EU) 2016/679 of the European Parliament and of the Council, document 32016R0679 (April 27, 2016), https://eur-lex.europa.eu/eli/reg/2016/679/oj.

144 more than 95,000 complaints: Samuel Stolton, "95,000 Complaints Issued to EU Data Protection Authorities," EURACTIV, January 28, 2019, https://www.euractiv.com/section/data-protection/news/95000-complaints-issued-to-eu-data-protection-authorities/.

145 "We're still nailing down": David Ingram and Joseph Menn, "Exclusive: Facebook CEO stops short of extending European privacy globally," Reuters, April 3, 2018, https://www.reuters.com/article/us-facebook-ceo-privacy-exclusive/exclusive-facebook-ceo-stops-short-of-extending-european-privacy-globally-idUSKCN1HA2M1.

145 under significant criticism: Josh Constine, "Zuckerberg says Facebook will offer GDPR privacy controls everywhere," Techcrunch, April 4, 2018, https://techcrunch.com/2018/04/04/zuckerberg-gdpr/.

145 "If people really knew": Nicholas Confessore, "The Unlikely Activists Who Took On Silicon Valley—and Won," *New York Times*, August 14, 2018, https://www.nytimes.com/2018/08/14/magazine/facebook-google-privacy-data.html.

149 Nudges are design features: Richard H. Thaler and Cass R. Sunstein, *Nudge: Improving Decisions About Health, Wealth and Happiness* (New Haven: Yale University Press, 2008).

149   a neutral online identifier: Jordan Mitchell, "The Evolution of the Internet, Identity, Privacy and Tracking," IAB Technology Laboratory, September 4, 2019, https://iabtechlab.com/blog/evolution-of-internet-identity-privacy-tracking/.

150   "low-tech, defensive": Peter Maass, "Your FTC Privacy Watchdogs: Low-Tech, Defensive, Toothless," *Wired*, June 28, 2012, https://www.wired.com/2012/06/ftc-fail/.

151   Senator Kirsten Gillibrand: Lauren Feiner, "Sen. Gillibrand Proposes a New Government Agency to Protect Privacy on the Internet," CNBC, February 13, 2020, https://www.cnbc.com/2020/02/12/gillibrand-unveils-another-privacy-proposal-with-new-agency.html.

CHAPTER 6: Can Humans Flourish in a World of Smart Machines?

153   a 142-mile car race: Joseph Hooper, "From Darpa Grand Challenge 2004DARPA's Debacle in the Desert," *Popular Science*, June 4, 2004, https://www.popsci.com/scitech/article/2004-06/darpa-grand-challenge-2004darpas-debacle-desert/.

154   Stanley completed the course: David Orenstein, "Stanford Team's Win in Robot Car Race Nets $2 Million Prize," Stanford News Service, October 11, 2005, http://news.stanford.edu/news/2005/october12/stanleyfinish-100905.html.

154   "It's a no-brainer": Joan Robinson, "Robotic Vehicle Wins Race Under Team Leader Sebastian Thrun," Springer, November 8, 2005, http://www.springer.com/about+springer/media/pressreleases?SGWID=0-11002-2-803827-0.

154   California alone licensed: Raymond Perrault et al., *Artificial Intelligence Index Report 2019*, AI Index Steering Committee, Human-Centered AI Institute, Stanford University, December 2019, 129–31, https://euagenda.eu/upload/publications/untitled-283856-ea.pdf.

154   The World Health Organization estimates: "Road Safety," World Health Organization, https://www.who.int/data/maternal-newborn-child-adolescent/monitor.

154   more than 90 percent: National Highway Traffic Safety Administration, "Traffic Safety Facts: 2017 Data," US Department of Transportation, May 2019, https://crashstats.nhtsa.dot.gov/Api/Public/ViewPublication/812687.

154   making commuting time more productive: Peter Diamandis, "Self-Driving Cars Are Coming," *Forbes*, August 13, 2014, https://www.forbes.com/sites/peterdiamandis/2014/10/13/self-driving-cars-are-coming/.

155   Accordingly, most participants: Jean-François Bonnefon, Azim Shariff, and Iyad Rahwan, "The Social Dilemma of Autonomous Vehicles," *Science* 352, no. 6293 (June 24, 2016): 1573–76, https://doi.org/10.1126/science.aaf2654.

156   IBM's Deep Blue computer had "unseated humanity": Bruce Weber, "Swift and Slashing, Computer Topples Kasparov," *New York Times*, May 12, 1997, https://www.nytimes.com/1997/05/12/nyregion/swift-and-slashing-computer-topples-kasparov.html.

157   "from another dimension": Dawn Chan, "The AI That Has Nothing to Learn from Humans," *Atlantic*, October 20, 2017, https://www.theatlantic.com/technology/archive/2017/10/alphago-zero-the-ai-that-taught-itself-go/543450/.

158   50 percent of jobs in the American economy: Carolyn Dimitri, Anne Effland, and Neilson Conklin, "The 20th Century Transformation of U.S. Agriculture and Farm Policy," Economic Information Bulletin Number 3, June 2005, https://www.ers.usda.gov/webdocs/publications/44197/13566_eib3_1_.pdf.

158   digital automation can replace: Nick Bostrom and Eliezer Yudkowsky, "The Ethics of Artificial Intelligence," in *Cambridge Handbook of Artificial Intelligence*, edited by Keith Frankish and William M. Ramsey (Cambridge, UK: Cambridge University Press, 2014), 316–34.

159   Will AGI put humanity: Edward Feigenbaum et al., *Advanced Software Applications in Japan* (Park Ridge, NJ: Noyes Data Corporation, 1995).

161   problems in reasoning have the potential: Yaniv Taigman et al., "DeepFace: Closing the Gap to Human-Level Performance in Face Verification," *2014 IEEE Conference on Computer Vision and Pattern Recognition (CVPR 2014)* (New York: IEEE, 2014), 1701–8, https://doi.org/10.1109/CVPR.2014.220.

162   "nine-layer deep neural network": Ibid.

163   Anyone using Zoom videoconferencing: "Language Interpretation in Meetings and Webinars," Zoom Help Center, https://support.zoom.us/hc/en-us/articles/360034919791-Language-interpretation-in-meetings-and-webinars.

164   "artificial intelligence (AI) system": Scott Mayer McKinney et al., "International Evaluation of an AI System for Breast Cancer Screening," *Nature* 577 (January 2020): 89–94, https://doi.org/10.1038/s41586-019-1799-6.

164   "an algorithm that can detect": Pranav Rajpurkar et al., "Radiologist-Level Pneumonia Detection on Chest X-Rays with Deep Learning," CheXNet, December 25, 2017, http://arxiv.org/abs/1711.05225.

164   "people should stop training radiologists": Geoff Hinton, "Geoff Hinton: On Radiology," Creative Destruction Lab, uploaded to YouTube November 24, 2016, https://www.youtube.com/watch?v=2HMPRXstSvQ.

164   the work radiologists and other medical professionals do: Hugh Harvey, "Why AI Will Not Replace Radiologists," Medium, April 7, 2018, https://towardsdatascience.com/why-ai-will-not-replace-radiologists-c7736f2c7d80.

165   "deep learning models to be equivalent": Xiaoxuan Liu et al., "A Comparison of Deep Learning Performance Against Health-Care Professionals in Detecting

Diseases from Medical Imaging: A Systematic Review and Meta-Analysis," *Lancet Digital Health* 1, no. 6 (October 1, 2019): e271–97, https://doi.org/10.1016/S2589-7500(19)30123-2.

166   eighty distinct AI ethics documents: Anna Jobin, Marcello Ienca, and Effy Vayena, "The Global Landscape of AI Ethics Guidelines," *Nature Machine Intelligence* 1, no. 9 (September 2019): 389–99, https://doi.org/10.1038/s42256-019-0088-2.

167   "Everyone wants to have a happy life": Wagner James Au, "VR Will Make Life Better—or Just Be an Opiate for the Masses," *Wired*, February 25, 2016, https://www.wired.com/2016/02/vr-moral-imperative-or-opiate-of-masses/.

168   *no one* would plug into the experience machine: Robert Nozick, *The Examined Life: Philosophical Meditations* (New York: Simon & Schuster, 2006), 106.

168   "importantly connected to reality": Ibid.

169   "gradually degrade the ways": Jaron Lanier, *You Are Not a Gadget: A Manifesto* (New York: Knopf Doubleday Publishing Group, 2010), x.

170   economist Gregory Clark: Gregory Clark, *A Farewell to Alms: A Brief Economic History of the World* (Princeton: Princeton University Press, 2007), 1.

170   rising levels of wealth: Angus Deaton, *The Great Escape: Health, Wealth, and the Origins of Inequality* (Princeton: Princeton University Press, 2015).

172   "greatest public-health achievement": Adrienne LaFrance, "Self-Driving Cars Could Save Tens of Millions of Lives This Century," *Atlantic*, September 29, 2015, https://www.theatlantic.com/technology/archive/2015/09/self-driving-cars-could-save-300000-lives-per-decade-in-america/407956/.

172   Amartya Sen took up: Amartya Sen, *Development as Freedom* (New York: Anchor, 2000).

172   "wealth is clearly not the good": Aristotle, *Nicomachean Ethics*, trans. Roger Crisp (Cambridge, UK: Cambridge University Press, 2004), 7.

173   quality of people's lives: Our World in Data (website), https://ourworldindata.org.

174   displace the need for human labor: Carl Benedikt Frey and Michael A. Osborne, "The Future of Employment," *Technological Forecasting and Social Change* 114 (January 2017): 254–80, https://doi.org/10.1016/j.techfore.2016.08.019.

174   only 9 percent of jobs are truly threatened: "Automation and Independent Work in a Digital Economy," Organisation for Economic Co-operation and Development, May 2016, https://www.oecd.org/els/emp/Policy%20brief%20-%20Automation%20and%20Independent%20Work%20in%20a%20Digital%20Economy.pdf.

174   *"technological unemployment"*: John Maynard Keynes, "Economic Possibilities for Our Grandchildren (1930)," in *Essays in Persuasion* (New York: W. W. Norton & Company, 1963), 358–83.

174   Harvard economist Wassily Leontief: Charlotte Curtis, "Machines vs. Workers," *New York Times*, February 8, 1983, https://www.nytimes.com/1983/02/08 /arts/machines-vs-workers.html.

174   Hal Varian: Aaron Smith and Janna Anderson, "AI, Robotics, and the Future of Jobs," Pew Research Center, August 6, 2014, https://www.pewresearch.org /internet/2014/08/06/future-of-jobs/.

174   dentists and clergy remain relatively safe: Ibid.

175   More than 40 percent: Mark Fahey, "Driverless Cars Will Kill the Most Jobs in Select US States," CNBC, September 2, 2016, https://www.cnbc.com/2016/09 /02/driverless-cars-will-kill-the-most-jobs-in-select-us-states.html.

175   automation may have beneficial effects: Daron Acemoglu and Pascual Restrepo, "Artificial Intelligence, Automation and Work" (NBER Working Paper Series Working Paper 24196, January 2018), 43.

175   changes in the industrial use of robots: Daron Acemoglu and Pascual Restrepo, "Robots and Jobs" (NBER Working Paper Series, Working Paper 23285, March 2017), https://www.nber.org/system/files/working_papers/w23285 /w23285.pdf.

176   risks of automation are unevenly distributed: Jason Furman, "Is This Time Different? The Opportunities and Challenges of Artificial Intelligence," remarks at AI Now: The Social and Economic Implications of Artificial Intelligence Technologies in the Near Term, New York University, New York, NY, July 7, 2016, https://obamawhitehouse.archives.gov/sites/default/files/page/files/2016 0707_cea_ai_furman.pdf.

177   open letter on behalf of AI: Stuart Russell, "Open Letter on AI," *Berkeley Engineer*, January 15, 2015, https://engineering.berkeley.edu/news/2015/11/open-letter -on-ai/.

177   "third revolution in warfare": Stuart Russell, "Take a Stand on AI Weapons," in "Robotics: Ethics of Artificial Intelligence," *Nature* 521, no. 7553 (May 17, 2015): 415–18, https://www.nature.com/news/robotics-ethics-of-artificial -intelligence-1.17611#/russell.

177   "Autonomous weapons are ideal": Russell, "Open Letter on AI."

177   "develop, manufacture, trade": "Lethal Autonomous Weapons Pledge," Future of Life Institute, 2018, https://futureoflife.org/lethal-autonomous-weapons-pledge/.

178   human *in* the loop: Daron Acemoglu and Pascual Restrepo, "The Wrong Kind of AI?," *TNIT News*, special issue, December 2018, https://idei.fr/sites/default /files/IDEI/documents/tnit/newsletter/newsletter_tnit_2019.pdf.

179   an effective subsidy to corporations: Daron Acemoglu, *Redesigning AI: Work, Democracy, and Justice in the Age of Automation* (Cambridge: MIT Press, 2021).

179   "society-in-the-loop": Iyad Rahwan, "Society-in-the-Loop," *Ethics and Information Technology* 20, no. 1 (March 2018): 5–14, https://link.springer.com/article/10.1007/s10676-017-9430-8.

180   amplifying the voices of workers: "Worker Voices," New America, November 21, 2019, http://newamerica.org/work-workers-technology/events/worker-voices/.

180   Darren Walker: Steven Greenhouse, "Where Are the Workers When We Talk About the Future of Work?," *American Prospect*, October 22, 2019, https://prospect.org/labor/where-are-the-workers-when-we-talk-about-the-future-of-work/.

180   it was a union that secured: Adam Seth Litwin, "Technological Change at Work," *ILR Review* 64, no. 5 (October 1, 2011): 863–88, https://doi.org/10.1177/001979391 106400502.

180   A movement is also afoot: Alana Semuels, "Getting Rid of Bosses," *Atlantic*, July 8, 2015, https://www.theatlantic.com/business/archive/2015/07/no-bosses-worker-owned-cooperatives/397007/.

180   what else might workers demand?: Pegah Moradi and Karen Levy, "The Future of Work in the Age of AI: Displacement or Risk-Shifting?," in *Oxford Handbook of Ethics of AI*, edited by Marus D. Dubber, Frank Pasquale, and Sunit Das (Oxford: Oxford University Press, 2020), 4–5.

181   the proper role of a corporation: "Business Roundtable Redefines the Purpose of a Corporation to Promote 'An Economy That Serves All Americans,'" Business Roundtable, August 19, 2019, https://www.businessroundtable.org/business-roundtable-redefines-the-purpose-of-a-corporation-to-promote-an-economy-that-serves-all-americans.

181   "accountable capitalism": "Empowering Workers Through Accountable Capitalism," Warren Democrats, 2020, https://elizabethwarren.com/plans/accountable-capitalism.

182   a new engine of productivity and growth: Daron Acemoglu and Pascual Restrepo, "The Wrong Kind of AI? Artificial Intelligence and the Future of Labor Demand," *Cambridge Journal of Regions, Economy and Society* 13, no. 1 (November 2019): 25–35.

182   "The big trap": Abby Vesoulis, "This Presidential Candidate Wants to Give Every Adult $1,000 a Month," *Time*, February 13, 2019, https://time.com/5528621/andrew-yang-universal-basic-income/.

183   "Right now, the human worker": Kevin J. Delaney, "The Robot That Takes Your Job Should Pay Taxes, Says Bill Gates," Quartz, February 17, 2017, https://qz.com/911968/bill-gates-the-robot-that-takes-your-job-should-pay-taxes/.

184   the push for UBI seems a bit misplaced: Dylan Matthews, "Andrew Yang's Basic

Income Can't Do Enough to Help Workers Displaced by Technology," Vox, October 18, 2019, https://www.vox.com/future-perfect/2019/10/18/20919322 /basic-income-freedom-dividend-andrew-yang-automation.

184   Jason Furman: Furman, "Is This Time Different?"

184   donate any windfall profits from AI: Cullen O'Keefe et al., "The Windfall Clause: Distributing the Benefits of AI for the Common Good," Cornell University, January 24, 2020, http://arxiv.org/abs/1912.11595.

185   18.7 percent of GDP on social programs: "Social Spending," Organisation for Economic Co-operation and Development, http://data.oecd.org/socialexp /social-spending.htm.

185   devotes far more resources to social spending: Ana Swanson, "How the U.S. Spends More Helping Its Citizens than Other Rich Countries, but Gets Way Less," *Washington Post*, April 9, 2015, https://www.washingtonpost.com/news /wonk/wp/2015/04/09/how-the-u-s-spends-more-helping-its-citizens-than -other-rich-countries-but-gets-way-less/.

185   its rate of premature deaths: Jacob Funk Kirkegaard, "The True Levels of Government and Social Expenditures in Advanced Economies," Peterson Institute for International Economics, Policy Brief 15-4, March 2015, https://piie.com /publications/pb/pb15-4.pdf, 19.

185   twice as likely to move to the top 20 percent: "Americans Overestimate Social Mobility in Their Country," *Economist*, February 14, 2018, https://www .economist.com/graphic-detail/2018/02/14/americans-overestimate-social -mobility-in-their-country.

185   Harvard economist Raj Chetty: Ezra Klein, "You Have a Better Chance of Achieving 'the American Dream' in Canada than in America," Vox, August 15, 2019, https://www.vox.com/2019/8/15/20801907/raj-chetty-ezra-klein-social -mobility-opportunity.

CHAPTER 7: Will Free Speech Survive the Internet?

187   Twitter CEO Jack Dorsey received an urgent call: Kate Conger and Mike Isaac, "Inside Twitter's Decision to Cut Off Trump," *New York Times*, January 16, 2021, https://www.nytimes.com/2021/01/16/technology/twitter-donald -trump-jack-dorsey.html.

187   twelve-hour suspension on his account: Haley Messenger, "Twitter to Uphold Permanent Ban Against Trump, Even If He Were to Run for Office Again," NBC News, February 10, 2021, https://www.nbcnews.com/news/amp/ncna 1257269.

187 "the President's statements": "Permanent Suspension of @realDonaldTrump," *Twitter* (blog), January 8, 2021, https://blog.twitter.com/en_us/topics/company /2020/suspension.html.

187 "Offline harm": Jack Dorsey (@jack), "I believe this was the right decision for Twitter. We faced an extraordinary and untenable circumstance, forcing us to focus all of our actions on public safety. Offline harm as a result of online speech is demonstrably real, and what drives our policy and enforcement above all," Twitter, January 13, 2021, https://twitter.com/jack/status /1349510770992640001.

189 Facebook is more like a government: Franklin Foer, "Facebook's War on Free Will," *Guardian*, September 19, 2017, http://www.theguardian.com/technology /2017/sep/19/facebooks-war-on-free-will

189 terrorist attack on mosques in Christchurch: Jon Porter, "Facebook Says the Christchurch Attack Live Stream Was Viewed by Fewer than 200 People," *Verge*, March 19, 2019, https://www.theverge.com/2019/3/19/18272342/facebook -christchurch-terrorist-attack-views-report-takedown.

190 An investigation by ProPublica: Ryan McCarthy, "'Outright Lies': Voting Misinformation Flourishes on Facebook," ProPublica, July 16, 2020, https:// www.propublica.org/article/outright-lies-voting-misinformation-flourishes-on -facebook.

191 "to provide redress": Gurbir S. Grewal et al., Attorneys General letter to Mark Zuckerberg and Sheryl Sandberg, August 5, 2020, https://www.nj.gov/oag /newsreleases20/AGs-Letter-to-Facebook.pdf.

191 Keith Rabois: June Cohen, "Rabois' Comments on 'Faggots' Derided Across University," *Stanford Daily*, February 6, 1992, https://archives.stanforddaily.com /1992/02/06?page=1&section=MODSMD_ARTICLE5#article.

191 "expose these freshman ears": Keith Rabois, "Rabois: My Intention Was to Make a Provocative Statement," *Stanford Daily*, February 7, 1992, https:// archives.stanforddaily.com/1992/02/07?page=5&section=MODSMD_ARTICLE 21#article.

192 "This vicious tirade": "Officials Condemn Homophobic Incident; No Prosecution Planned," Stanford News Service, February 12, 1992, https://news.stanford .edu/pr/92/920212Arc2432.html.

193 "generated more total engagement": Craig Silverman, "This Analysis Shows How Viral Fake Election News Stories Outperformed Real News on Facebook," BuzzFeed News, November 16, 2016, https://www.buzzfeednews.com/article /craigsilverman/viral-fake-election-news-outperformed-real-news-on-facebook.

193 believe the story about Gates: Ciara O'Rourke, "No, the Gates Foundation Isn't Pushing Microchips with All Medical Procedures," PolitiFact, May 20,

2020, https://www.politifact.com/factchecks/2020/may/20/facebook-posts/no
-gates-foundation-isnt-pushing-microchips-all-me/; Linley Sanders, "The Difference Between What Republicans and Democrats Believe to Be True About
COVID-19," YouGov, May 26, 2020, https://today.yougov.com/topics/politics
/articles-reports/2020/05/26/republicans-democrats-misinformation.

193   When Gates was asked: Steven Levy, "Bill Gates on Covid: Most US Tests Are
'Completely Garbage,'" *Wired*, August 7, 2020, https://www.wired.com/story
/bill-gates-on-covid-most-us-tests-are-completely-garbage/.

193   "putting power in people's hands": Mark Zuckerberg, "A Blueprint for Content
Governance and Enforcement," Facebook, November 15, 2018, https://www
.facebook.com/notes/751449002072082/.

194   Billions of pieces: Mark Zuckerberg, "Building Global Community," Facebook,
February 16, 2017, https://www.facebook.com/notes/mark-zuckerberg/building
-global-community/10154544292806634/.

194   "over 300 hours of video": "YouTube Jobs," YouTube, https://www.youtube.com
/jobs/.

194   more than 500 million tweets: Raffi Krikorian, "New Tweets per Second Record,
and How!," *Twitter* (blog), August 16, 2013, https://blog.twitter.com/engineering
/en_us/a/2013/new-tweets-per-second-record-and-how.html.

195   the print version was discontinued: Christine Kearney, "Encyclopaedia Britannica: After 244 Years in Print, Only Digital Copies Sold," *Christian Science
Monitor*, March 14, 2012, https://www.csmonitor.com/Business/Latest-News
-Wires/2012/0314/Encyclopaedia-Britannica-After-244-years-in-print-only
-digital-copies-sold.

195   more than 3.9 million articles: Camille Slater, "Wikipedia vs Britannica: A
Comparison Between Both Encyclopedias," SciVenue, November 17, 2017,
http://scivenue.com/2017/11/17/wikipedia-vs-britannica-encyclopedia/.

195   more than 6.2 million articles: Wikipedia, s.v., "Wikipedia: Statistics," https://
en.wikipedia.org/wiki/Wikipedia:Statistics; "Wikistats: Statistics for Wikimedia Projects," Wikimedia Statistics, https://stats.wikimedia.org/#/en.wikipedia
.org/reading/total-page-views/normal%7Cbar%7C2-year%7C~total%7C
monthly.

195   two-thirds of visits: Wikipedia, s.v., "Google Statistics," May 1, 2013, https://
en.wikipedia.org/w/index.php?title=Wikipedia:Google_statistics&oldid
=553012140.

196   spending on SEO: Greg Sterling, "Forecast Says SEO-Related Spending Will
Be Worth $80 Billion by 2020," Search Engine Land, April 19, 2016, https://
searchengineland.com/forecast-says-seo-related-spending-will-worth-80
-billion-2020-247712.

198   Thomas Jefferson: Thomas Jefferson, "Letter to Charles Yancey," Manuscript /Mixed Material, January 6, 1816, https://www.loc.gov/resource/mtj1.048_0731 _0734.

198   "To refuse a hearing": John Stuart Mill, *Utilitarianism and On Liberty: Including Mill's 'Essay on Bentham' and Selections from the Writings of Jeremy Bentham and John Austin*, ed. Mary Warnock, 2nd ed, (Hoboken: Wiley-Blackwell, 2003), 100.

199   freedom to choose their own: Ibid, 134.

199   Denying freedom of speech: Ibid., 100.

199   Supreme Court Justice Louis Brandeis: Louis Brandeis statement, *Whitney v. California*, 274 U.S. 357 (1927), https://www.law.cornell.edu/supremecourt/text /274/357.

200   "single biggest threat to democracy": Jeffrey Goldberg, "Why Obama Fears for Our Democracy," *Atlantic*, November 16, 2020, https://www.theatlantic.com /ideas/archive/2020/11/why-obama-fears-for-our-democracy/617087/.

201   the Supreme Court permits limits: *Chaplinsky v. State of New Hampshire*, 315 U.S. 568, Supreme Court of the United States (1942), https://www.law.cornell.edu /supremecourt/text/315/568.

201   "time to stop shitposting": Raymond Lin, "New Zealand Shooter Kills 50 in Attack on Mosques," Guide Post Daily, April 1, 2019, https://gnnguidepost.org/2923 /news/new-zealand-shooter-kills-50-in-attack-on-mosques/.

202   7 or 8 out of every 10,000: Guy Rosen, "Community Standards Enforcement Report, Fourth Quarter 2020," Facebook Newsroom, February 11, 2021, https://about .fb.com/news/2021/02/community-standards-enforcement-report-q4-2020/.

203   "breeding extremism": Cass Sunstein, *Republic.com 2.0* (Princeton: Princeton University Press, 2007), 69, 78

203   "asking an epidemiologist": Joshua Cohen, "Against Cyber-Utopianism," *Boston Review*, June 19, 2012, https://bostonreview.net/joshua-cohen-reflections-on -information-technology-and-democracy.

203   this area of scholarly research: Nathaniel Persily and Joshua A. Tucker, eds., *Social Media and Democracy: The State of the Field, Prospects for Reform*, SSRC Anxieties of Democracy (Cambridge, UK: Cambridge University Press, 2020).

203   "cross-cutting interactions are more frequent": Pablo Barberá, "Social Media, Echo Chambers, and Political Polarization," in *Social Media and Democracy: The State of the Field, Prospects for Reform*, edited by Nathaniel Persily and Joshua A. Tucker, SSRC Anxieties of Democracy (Cambridge, UK: Cambridge University Press, 2020), 34–55.

204   a small number of very frequent users: Pablo Barberá and Gonzalo Rivero,

"Understanding the Political Representativeness of Twitter Users," *Social Science Computer Review* 33, no. 6 (2015): 712–29, https://doi.org/10.1177/0894439314558836.

204 rely primarily on centrist websites: Andrew M. Guess, Brendan Nyhan, and Jason Reifler, "Exposure to Untrustworthy Websites in the 2016 US Election," *Nature Human Behaviour* 4, no. 5 (March 2, 2020): 472–80, https://doi.org/10.1038/s41562-020-0833-x.

204 And approximately 30 percent: E. Bakshy, S. Messing, and L. A. Adamic, "Exposure to Ideologically Diverse News and Opinion on Facebook," *Science* 348, no. 6239 (June 5, 2015): 1130–32, https://doi.org/10.1126/science.aaa1160.

204 most people are exposed: Matthew Barnidge, "Exposure to Political Disagreement in Social Media Versus Face-to-Face and Anonymous Online Settings," *Political Communication* 34, no. 2 (April 3, 2017): 302–21, https://doi.org/10.1080/10584609.2016.1235639.

204 people in the oldest age cohorts: Levi Boxell, Matthew Gentzkow, and Jesse M. Shapiro, "Greater Internet Use Is Not Associated with Faster Growth in Political Polarization Among US Demographic Groups," *Proceedings of the National Academy of Sciences of the United States of America* 114, no. 40 (October 3, 2017): 10612–17, https://doi.org/10.1073/pnas.1706588114.

204 social media usage and polarization: Hunt Allcott et al., "The Welfare Effects of Social Media," *American Economic Review* 110, no. 3 (March 1, 2020): 629–76, https://doi.org/10.1257/aer.20190658.

205 "claims that contradict": Andrew M. Guess and Benjamin A. Lyons, "Misinformation, Disinformation, and Online Propaganda," in *Social Media and Democracy: The State of the Field, Prospects for Reform*, edited by Nathaniel Persily and Joshua A. Tucker, SSRC Anxieties of Democracy (Cambridge, UK: Cambridge University Press, 2020), 10–33.

205 fifty comments daily: Andrew Dawson and Martin Innes, "How Russia's Internet Research Agency Built Its Disinformation Campaign," *Political Quarterly* 90, no. 2 (June 2019): 245–56, https://doi.org/10.1111/1467-923X.12690.

206 tracked more than 2,500 adults: Guess, Nyhan, and Reifler, "Exposure to Untrustworthy Websites in the 2016 US Election."

206 older people are much more likely: Andrew Guess, Jonathan Nagler, and Joshua Tucker, "Less than You Think: Prevalence and Predictors of Fake News Dissemination on Facebook," *Science Advances* 5, no. 1 (January 2019), https://doi.org/10.1126/sciadv.aau4586.

206 real material from local news sources: Jun Yin et al., "Social Spammer Detection: A Multi-Relational Embedding Approach," in *Advances in Knowledge Discovery* *and Data Mining,* Pacific-Asia Conference on Knowledge Discovery and Data Mining, edited by Hady W. Lauw et al. (Melbourne: Springer Nature, 2018), 615–27, https://doi.org/10.1007/978-3-319-93034-3_49.

206   committed to their previous misconceptions: Brendan Nyhan and Jason Reifler, "When Corrections Fail: The Persistence of Political Misperceptions," *Political Behavior* 32, no. 2 (June 1, 2010): 303–30, https://doi.org/10.1007/s11109-010 -9112-2.

207   a misinformed citizenry can be quite dangerous: James H. Kuklinski et al., "Misinformation and the Currency of Democratic Citizenship," *Journal of Politics* 62, no. 3 (August 1, 2000): 790–816, https://doi.org/10.1111/0022-3816.00033.

207   hate speech has found: Alexandra A. Siegel, "Online Hate Speech," in *Social Media and Democracy: The State of the Field, Prospects for Reform,* edited by Joshua A. Tucker and Nathaniel Persily, SSRC Anxieties of Democracy (Cambridge, UK: Cambridge University Press, 2020), 56–88.

207   analyzed more than 750 million tweets: Alexandra Siegel et al., "Trumping Hate on Twitter? Online Hate in the 2016 US Election and Its Aftermath," March 6, 2019, https://smappnyu.wpcomstaging.com/wp-content/uploads /2019/04/US_Election_Hate_Speech_2019_03_website.pdf.

207   Facebook claims to take down: Laura W. Murphy, "Facebook's Civil Rights Audit—Final Report," Facebook Newsroom, July 8, 2020, https://about.fb.com /wp-content/uploads/2020/07/Civil-Rights-Audit-Final-Report.pdf.

207   Exposure was also high: James Hawdon, Atte Oksanen, and Pekka Räsänen, "Exposure to Online Hate in Four Nations: A Cross-National Consideration," *Deviant Behavior* 38, no. 3 (March 4, 2017): 254–66, https://doi.org/10.1080/0 1639625.2016.1196985.

207   a small but energetic minority: Manoel Horta Ribeiro et al., "Characterizing and Detecting Hateful Users on Twitter," arXiv, March 23, 2018, https://arxiv.org/abs /1803.08977v1.

207   softer and more indirect posts: Nick Beauchamp, Ioana Panaitiu, and Spencer Piston, "Trajectories of Hate: Mapping Individual Racism and Misogyny on Twitter" (unpublished working paper, 2018).

208   it could accurately predict: Walid Magdy et al., "#ISISisNotIslam or #DeportAllMuslims? Predicting Unspoken Views," in *Proceedings of the 8th ACM Conference on Web Science,* WebSci '16 (Hannover, Germany: Association for Computing Machinery, 2016), 95–106, https://doi.org/10.1145/2908131 .2908150.

208   One study in Germany: Karsten Müller and Carlo Schwarz, "Fanning the Flames of Hate: Social Media and Hate Crime," *Journal of the European Economic Association,* October 30, 2020, doi.org/10.1093/jeea/jvaa045.

208   Likewise, in the United States: Karsten Müller and Carlo Schwarz, "From Hashtag to Hate Crime: Twitter and Anti-Minority Sentiment" (unpublished working paper, 2020), https://papers.ssrn.com/abstract=3149103.

209   the core function of a platform: Tarleton Gillespie, *Custodians of the Internet: Platforms, Content Moderation, and the Hidden Decisions That Shape Social Media* (New Haven: Yale University Press, 2018).

209   Mark Zuckerberg wrote: Zuckerberg, "Building Global Community."

210   Community Standards Enforcement Report: Rosen, "Community Standards Enforcement Report, Fourth Quarter 2020."

210   According to the February 2021 report: Ibid.

210   "because bullying and harassment": "Facebook's Response to Australian Government Consultation on a New Online Safety Act," Facebook Newsroom, February 19, 2020, https://about.fb.com/wp-content/uploads/2020/02/Facebook-response-to-consultation-new-Online-Safety-Act.pdf.

210   YouTube CEO Susan Wojcicki: Susan Wojcicki, "Expanding Our Work Against Abuse of Our Platform," *YouTube Official Blog*, December 5, 2017, https://blog.youtube/news-and-events/expanding-our-work-against-abuse-of-our/.

211   "the team responsible": Zuckerberg, "A Blueprint for Content Governance and Enforcement."

211   "distressing videos and photographs": Sandra E. Garcia, "Ex–Content Moderator Sues Facebook, Saying Violent Images Caused Her PTSD," *New York Times*, September 25, 2018, https://www.nytimes.com/2018/09/25/technology/facebook-moderator-job-ptsd-lawsuit.html. See also the pioneering ethnographic work on this topic by Sarah T. Roberts, *Behind the Screen: Content Moderation in the Shadows of Social Media* (New Haven: Yale University Press, 2019).

211   "Facebook had failed": Casey Newton, "Facebook Will Pay $52 Million in Settlement with Moderators Who Developed PTSD on the Job," *Verge*, May 12, 2020, https://www.theverge.com/2020/5/12/21255870/facebook-content-moderator-settlement-scola-ptsd-mental-health.

211   has offered a taxonomy: Nathaniel Persily, "The Internet's Challenge to Democracy: Framing the Problem and Assessing Reforms," Kofi Annan Commission on Elections and Democracy in the Digital Age, March 4, 2019, https://pacscenter.stanford.edu/publication/the-internets-challenge-to-democracy-framing-the-problem-and-assessing-reforms/.

211   Zuckerberg discussed exactly that phenomenon: Mark Zuckerberg, "Preparing for Elections," Facebook, September 13, 2018, https://www.facebook.com/notes/mark-zuckerberg/preparing-for-elections/10156300047606634/.

212   "posts that are rated as false": Ibid.

213   "increasing our use of machine learning": Vijaya Gadde and Matt Derella, "An Update on Our Continuity Strategy During COVID-19," *Twitter* (blog), March 16, 2020, https://blog.twitter.com/en_us/topics/company/2020/An-update-on-our-continuity-strategy-during-COVID-19.html.

213   "we will temporarily start": The YouTube Team, "Protecting Our Extended Workforce and the Community," *YouTube Official Blog*, March 16, 2020, https://blog.youtube/news-and-events/protecting-our-extended-workforce-and/.

213   "This was an issue": Guy Rosen (@guyro), "We've restored all the posts that were incorrectly removed, which included posts on all topics—not just those related to COVID-19. This was an issue with an automated system that removes links to abusive websites, but incorrectly removed a lot of other posts too," Twitter, March 17, 2020, https://twitter.com/guyro/status/1240088303497400320?lang=en.

214   the new private governors of speech: Kate Klonick, "The New Governors: The People, Rules, and Processes Governing Online Speech," *Harvard Law Review* 131 (2018): 1598–670, https://harvardlawreview.org/wp-content/uploads/2018/04/1598-1670_Online.pdf. See also Klonick, "The Facebook Oversight Board: Creating an Independent Institution to Adjudicate Online Free Expression," *Yale Law Journal* 129, no. 8 (2020): 2418–99, https://papers.ssrn.com/sol3/papers.cfm?abstract_id=3639234; Evelyn Douek, "Facebook's 'Oversight Board:' Move Fast with Stable Infrastructure and Humility," *North Carolina Journal of Law & Technology* 21, no. 1 (2019): 1–78, https://papers.ssrn.com/sol3/papers.cfm?abstract_id=3365358.

214   "I've increasingly come to believe": Zuckerberg, "A Blueprint for Content Governance and Enforcement."

216   "co-opted in some shape or form": Ben Smith, "Trump Wants Back on Facebook. This Star-Studded Jury Might Let Him," *New York Times*, January 24, 2021, https://www.nytimes.com/2021/01/24/business/media/trump-facebook-oversight-board.html.

217   "there might have to be some revengeance": Clarence Brandenburg, Appellant, v. State of Ohio, 395 U.S. 444 (June 9, 1969), Legal Information Institute, Cornell Law School, https://www.law.cornell.edu/supremecourt/text/395/444.

217   Racist and anti-Semitic rhetoric: Yascha Mounk, "Verboten: Germany's Risky Law for Stopping Hate Speech on Facebook and Twitter," *New Republic*, April 3, 2018, https://newrepublic.com/article/147364/verboten-germany-law-stopping-hate-speech-facebook-twitter.

218   a commitment to freedom of speech: Renee DiResta, "Free Speech Is Not the Same as Free Reach," *Wired*, August 30, 2018, https://www.wired.com/story/free-speech-is-not-the-same-as-free-reach/.

219   In places where hate speech laws: Joanna Plucinska, "Hate Speech Thrives Underground," *Politico*, February 7, 2018, https://www.politico.eu/article/hate-speech-and-terrorist-content-proliferate-on-web-beyond-eu-reach-experts/.

219   "defending the principal channels": Tim Wu, "Is the First Amendment Obsolete?," *Michigan Law Review* 117, no. 3 (2018): 547–81, https://michiganlawreview.org/wp-content/uploads/2018/12/117MichLRev547_Wu.pdf.

220   According to a 2021 Pew Survey: Pew Research Center, January 2021, "The State of Online Harassment."

220   Laura Bates founded: https://www.amnesty.org/en/latest/news/2017/11/amnesty-reveals-alarming-impact-of-online-abuse-against-women/.

220   not to guarantee high-quality public debate: Tim Wu, "Is the First Amendment Obsolete?," *Michigan Law Review* 117, no. 3 (2018): 547–81, https://michiganlawreview.org/wp-content/uploads/2018/12/117MichLRev547_Wu.pdf.

222   "liable as a publisher": Klonick, "The New Governors."

222   "to gain the benefits of editorial control": Ibid.

222   When considering something such as CDA 230: Daphne Keller, "Statement of Daphne Keller Before the United States Senate Committee on the Judiciary, Subcommittee on Intellectual Property, Hearing on the Digital Millennium Copyright Act at 22: How Other Countries Are Handling Online Piracy," March 10, 2020, https://www.judiciary.senate.gov/imo/media/doc/Keller%20Testimony.pdf.

222   "It is not merely an internet company": *New York Times* Editorial Board, "Joe Biden," *New York Times*, January 17, 2020, https://www.nytimes.com/interactive/2020/01/17/opinion/joe-biden-nytimes-interview.html.

223   Two additional things are different: Francis Fukuyama and Andrew Grotto, "Comparative Media Regulation in the United States and Europe," in *Social Media and Democracy: The State of the Field, Prospects for Reform*, edited by Joshua A. Tucker and Nathaniel Persily, SSRC Anxieties of Democracy (Cambridge, UK: Cambridge University Press, 2020), 199–219.

224   the migration to Parler: Stanford Internet Observatory, "Parler's First 13 Million Users," January 28, 2021, https://fsi.stanford.edu/news/sio-parler-contours.

227   Google is responsible for over 90 percent: Daisuke Wakabayashi and Tiffany Hsu, "Why Google Backtracked on Its New Search Results Look," *New York Times*, January 31, 2020, https://www.nytimes.com/2020/01/31/technology/google-search-results.html.

227   Facebook generates nearly 70 percent: "Social Media Stats Worldwide," StatCounter Global Stats, https://gs.statcounter.com/social-media-stats.

PART III: Recoding the Future

231   "Only if we acknowledge:" Sheila Jasanoff, *The Ethics of Invention: Technology and the Human Future* (New York: W. W. Norton & Company, 2016), 267.

CHAPTER 8: Can Democracies Rise to the Challenge?

233   "Due to our concerns": Alec Radford et al., "Better Language Models and Their Implications," OpenAI, February 14, 2019, https://openai.com/blog/better -language-models/.

234   Some AI scientists facetiously said: Yoav Goldberg (@yoavgo), "Just wanted to give you all a heads up, our lab found an amazing breakthrough in language understanding. but we also worry it may fall into the wrong hands. so we de- cided to scrap it and only publish the regular *ACL stuff instead. Big respect for the team for their great work," Twitter, February 15, 2019, https://twitter.com /yoavgo/status/1096471273050382337.

234   "humans find GPT-2 outputs convincing": Irene Solaiman, Jack Clark, and Miles Brundage, "GPT-2: 1.5B Release," OpenAI, November 5, 2019, https://openai .com/blog/gpt-2-1-5b-release/.

234   "extremist groups can use": Ibid.

235   the training data: Stephen Ornes, "Explainer: Understanding the Size of Data," Science News for Students, December 13, 2013, https://www.sciencenewsfor students.org/article/explainer-understanding-size-data.

235   *Kanye West Exclusive*: Arram Sabeti, "GPT-3," *Arram Sabeti* (blog), July 9, 2020, https://arr.am/2020/07/09/gpt-3-an-ai-thats-eerily-good-at-writing-almost -anything/.

236   "Why deep learning will never": Gwern Branwen, "GPT-3 Creative Fiction," gwern.net, June 19, 2020, https://www.gwern.net/GPT-3#why-deep-learning -will-never-truly-x; Kelsey Piper, "GPT-3, Explained: This New Language AI Is Uncanny, Funny—and a Big Deal," Vox, August 13, 2020, https://www.vox.com /future-perfect/21355768/gpt-3-ai-openai-turing-test-language.

238   trust in technology companies is declining: Carroll Doherty and Jocelyn Kiley, "Americans Have Become Much Less Positive About Tech Companies' Impact on the U.S.," Pew Research Center, July 29, 2019, https://www.pewresearch .org/fact-tank/2019/07/29/americans-have-become-much-less-positive-about -tech-companies-impact-on-the-u-s/; Ina Fried, "40% of Americans Believe Ar- tificial Intelligence Needs More Regulation," *Axios*, https://www.axios.com /big-tech-industry-global-trust-9b7c6c3c-98f1-4e80-8275-cf52446b1515.html.

238   Indeed, a growing number: Karen Hao, "The Coming War on the Hidden Algorithms That Trap People in Poverty," *MIT Technology Review*, December 4, 2020, https://www.technologyreview.com/2020/12/04/1013068/algorithms -create-a-poverty-trap-lawyers-fight-back/.

238 "ten arguments": Jaron Lanier, *Ten Arguments for Deleting Your Social Media Accounts Right Now* (New York: Henry Holt, 2018).

240 "count me as skeptical": Anand Giridharadas, "Deleting Facebook Won't Fix the Problem," *New York Times*, January 10, 2019, https://www.nytimes.com /2019/01/10/opinion/delete-facebook.html.

242 The platform brings together: "Where Do We Go as a Society?," vTaiwan, March 2016, https://info.vtaiwan.tw/.

242 More recently, Tang helped lead: Johns Hopkins, "Mortality Analyses," Johns Hopkins Coronavirus Resource Center, December 3, 2020, https://coronavirus .jhu.edu/data/mortality.

243 "Americans share every movement": Ezekiel J. Emanuel, Cathy Zhang, and Aaron Glickman, "Learning from Taiwan About Responding to Covid-19— and Using Electronic Health Records," STAT, June 30, 2020, https://www .statnews.com/2020/06/30/taiwan-lessons-fighting-covid-19-using-electronic -health-records/.

244 provide some important lessons: David J. Rothman, *Strangers at the Bedside: A History of How Law and Bioethics Transformed Medical Decision Making* (New York: Basic Books, 1991); Cathy Gere, *Pain, Pleasure, and the Greater Good* (Chicago: University of Chicago Press, 2017).

244 "an ethical tradition": Don Colburn, "Under Oath," *Washington Post*, October 22, 1991, http://www.washingtonpost.com/archive/lifestyle/wellness/1991/10/22 /under-oath/53407b39-4a27-4bca-91fe-44602fc05bbf/.

245 Those medical boards remain: Abraham Flexner, *Medical Education in the United States and Canada: A Report to the Carnegie Foundation for the Advancement of Teaching* (Boston: Merrymount Press, 1910), http://archive.org/details/medical education00flexiala.

247 "The licensing of software engineers": John Knight et al., "ACM Task Force on Licensing of Software Engineers Working on Safety-Critical Software," draft report, ACM, July 2000, http://kaner.com/pdfs/acmsafe.pdf.

248 Software Engineering Code of Ethics: Don Gotterbarn, Keith Miller, and Simon Rogerson, "Software Engineering Code of Ethics," *Communications of the ACM* 40, no. 11 (November 1, 1997): 110–18, https://doi.org/10.1145/265684.265699.

248 As Jack Dorsey lamented: Lauren Jackson and Desiree Ibekwe, "Jack Dorsey on Twitter's Mistakes," *New York Times*, August 19, 2020, https://www.nytimes .com/2020/08/07/podcasts/the-daily/Jack-dorsey-twitter-trump.html.

249   "Suppose somebody like Hitler": Michael Specter, "The Gene Hackers," *New Yorker*, November 8, 2015, https://www.newyorker.com/magazine/2015/11/16 /the-gene-hackers.

249   called for responsible publication guidelines: Rebecca Crootof, "Artificial Intelligence Research Needs Responsible Publication Norms," Lawfare, October 24, 2019, https://www.lawfareblog.com/artificial-intelligence-research-needs -responsible-publication-norms; Miles Brundage et al., *The Malicious Use of Artificial Intelligence: Forecasting, Prevention, and Mitigation* (Oxford: Future of Humanity Institute, 2018), https://maliciousaireport.com/.

250   An Israeli company: Faception (website), 2019, https://www.faception.com.

250   "the shape of the face": Mahdi Hashemi and Margeret Hall, "RETRACTED ARTICLE: Criminal Tendency Detection from Facial Images and the Gender Bias Effect," *Journal of Big Data* 7, no. 2 (January 7, 2020), https://journalofbigdata .springeropen.com/track/pdf/10.1186/s40537-019-0282-4.pdf.

250   such work is infected: Coalition for Critical Technology, "Abolish the #TechToPrisonPipeline," Medium, June 23, 2020, https://medium.com/@Coalition ForCriticalTechnology/abolish-the-techtoprisonpipeline-9b5b14366b16.

251   a new generation of civic-minded technologists: "About the Public Interest Technology University Network," New America, https://www.newamerica.org/pit /university-network/about-pitun/.

252   Margrethe Vestager: Natalia Drozdiak and Sam Schechner, "The Woman Who Is Reining In America's Technology Giants," *Wall Street Journal*, April 4, 2018, https://www.wsj.com/articles/the-woman-who-is-reining-in-americas-technology -giants-1522856428.

252   proposed new rules: "The Digital Markets Act: Ensuring Fair and Open Digital Markets," European Commission, https://ec.europa.eu/info/strategy/priorities -2019-2024/europe-fit-digital-age/digital-markets-act-ensuring-fair-and-open -digital-markets_en.

252   "their service providers": Kara Swisher, "White House. Red Chair. Obama Meets Swisher," Vox, February 15, 2015, https://www.vox.com/2015/2/15/11559056 /white-house-red-chair-obama-meets-swisher.

253   "It is better to buy": *Federal Trade Commission v. Facebook, Inc.*, United States District Court for the District of Columbia, December 9, 2020, https://www .ftc.gov/system/files/documents/cases/051_2021.01.21_revised_partially_redacted _complaint.pdf, 2.

253   A related lawsuit alleges: *The State of Texas, the State of Arkansas, the State of Idaho, the State of Indiana, the Commonwealth of Kentucky, the State of Mississippi, State of Missouri, State of North Dakota, State of South Dakota, and State of Utah vs. Google LLC*, United States District Court, Eastern District of Texas, Sher-

man Division, December 16, 2020, https://www.texasattorneygeneral.gov/sites
/default/files/images/admin/2020/Press/20201216%20COMPLAINT
_REDACTED.pdf.

254 "If we were to redo": Jackson and Ibekwe, "Jack Dorsey on Twitter's Mistakes."

254 "We have a responsibility": Nick Statt, "Mark Zuckerberg Apologizes for Facebook's Data Privacy Scandal in Full-Page Newspaper Ads," *Verge*, March 25,
2018, https://www.theverge.com/2018/3/25/17161398/facebook-mark-zuckerberg
-apology-cambridge-analytica-full-page-newspapers-ads.

254 "world that we actually ought": Eric Johnson, "Former Google Lawyer and Deputy U.S. CTO Nicole Wong on Recode Decode," Vox, September 12, 2018,
https://www.vox.com/2018/9/12/17848384/nicole-wong-cto-lawyer-google
-twitter-kara-swisher-decode-podcast-full-transcript.

254 "dieter who would rather": Joseph E. Stiglitz, "Meet the 'Change Agents' Who Are
Enabling Inequality," *New York Times*, August 20, 2018, https://www.nytimes
.com/2018/08/20/books/review/winners-take-all-anand-giridharadas.html.

255 has admitted the error of his ways: Edmund L. Andrews, "Greenspan Concedes
Error on Regulation," *New York Times*, October 23, 2008, https://www.nytimes
.com/2008/10/24/business/economy/24panel.html.

256 they include concrete legislative mandates: Accountable Capitalism Act, S. 3348,
115th Congress (2017–18), https://www.warren.senate.gov/imo/media/doc
/Accountable%20Capitalism%20Act%20One-Pager.pdf.

256 Elizabeth Warren's Accountable Capitalism Act: Elizabeth Warren, "Warren,
Carper, Baldwin, and Warner Form Corporate Governance Working Group to
Fundamentally Reform the 21st Century American Economy," Elizabeth Warren
(senate website), October 30, 2020, https://www.warren.senate.gov/newsroom
/press-releases/warren-carper-baldwin-and-warner-form-corporate-governance
-working-group-to-fundamentally-reform-the-21st-century-american-economy.

257 Many observers have speculated: Yoni Heisler, "What Ever Became of Microsoft's $150 Million Investment in Apple?," Engadget, May 20, 2014, https://
www.engadget.com/2014-05-20-what-ever-became-of-microsofts-150-million
-investment-in-apple.html.

257 softened its competitive stance: Jessica Bursztynsky, "Microsoft President: Being a Big Company Doesn't Mean You're a Monopoly," CNBC, September 10,
2019, https://www.cnbc.com/2019/09/10/microsoft-president-brad-smith-on
-facebook-and-google-antitrust-probes.html.

259 "You don't cut the future": Chris Mooney, "Requiem for an Office," *Bulletin of the
Atomic Scientists* 61, no. 5 (2005): 40–49, https://doi.org/10.2968/061005013, 45.

259 "stunning act of self-lobotomy": Ibid., 42.

259 "views on science and technology": Ibid., 45.

259 NOTA also incorporates: Sheila Jasanoff, *The Ethics of Invention: Technology and the Human Future* (New York: W. W. Norton, 2016).

260 "Nearly every national priority": "Tech Talent for 21st Century Government," The Partnership for Public Service, April 2020, https://ourpublicservice.org /wp-content/uploads/2020/04/Tech-Talent-for-21st-Century-Government.pdf, 2.

261 expertise to serve the public interest: Joseph J. Heck et al., *Inspired to Serve: The Final Report of the National Commission on Military, National, and Public Service*, March 2020, https://inspire2serve.gov/sites/default/files/final-report/Final%20 Report.pdf.

261 reinvention of the OTA: Derek Kilmer et al., "How Can Congress Work Better for the American People?," Select Committee on the Modernization of Congress, July 2019, https://modernizecongress.house.gov/final-report-116th; Elizabeth Fretwell et al., "Science and Technology Policy Assessment: A Congressionally Directed Review," National Academy of Public Administration, October 31, 2019, https://www.napawash.org/studies/academy-studies/science -and-technology-policy-assessment-for-the-us-congress; Office of Technology Assessment Improvement and Enhancement Act, H.R. 4426, 116th Congress (2019–2020), https://www.congress.gov/bill/116th-congress/house-bill/4426.

261 One priority for a revamped OTA: Jathan Sadowski, "The Much-Needed and Sane Congressional Office That Gingrich Killed Off and We Need Back," *Atlantic*, October 26, 2020, https://www.theatlantic.com/technology/archive /2012/10/the-much-needed-and-sane-congressional-office-that-gingrich-killed -off-and-we-need-back/264160/.

INDEX